高等职业教育机电类系列教材

SPECIFIED ENGLISH FOR CNC
2nd Edition

数控专业英语

第 2 版

主　编　王兆奇　刘向红
参　编　李晓会　黄桂芸　王　颖
主　审　王曙钊

机械工业出版社

本书共分 30 个单元，内容涉及数控发展历史、数控机床、数控技术的优点、现代数控机床、数控加工过程、数控编程、伺服控制、MCU 与 CPU、计算机数字控制、进给与转速、连续路径加工、加工中心的种类与组成、刀具监控及在程检测、切削刀具、刀具系统、自适应控制、电火花线切割加工、数控车床、数控铣床、数控机床主轴、计算机图形编程、CAD、CAM、CAD/CAM/CNC、柔性加工系统、数控职业范畴、工业机器人、计算机集成制造、面向 21 世纪的制造技术等，涵盖了数控发展的各个方向。此外，每篇文章后都附有专业词汇、课文注释、练习以及参考译文，一些课文后面还附有阅读材料供读者自学。

本书可作为高职高专数控技术、CAD、机电一体化等专业学生使用的专业英语教材，也可供有关工程技术人员参考。

图书在版编目（CIP）数据

数控专业英语/王兆奇，刘向红主编．—2 版．—北京：机械工业出版社，2012.8（2024.7 重印）
高等职业教育机电类系列教材
ISBN 978-7-111-38759-6

Ⅰ．①数… Ⅱ．①王…②刘… Ⅲ．数控技术—英语—高等职业教育—教材 Ⅳ．①H31

中国版本图书馆 CIP 数据核字（2012）第 145552 号

机械工业出版社（北京市百万庄大街 22 号　邮政编码 100037）
策划编辑：王英杰　　责任编辑：王英杰　刘良超　王晓燕
版式设计：霍永明　　责任校对：于新华
封面设计：鞠　杨　　责任印制：单爱军
北京虎彩文化传播有限公司印刷
2024 年 7 月第 2 版第 10 次印刷
184mm×260mm・11 印张・267 千字
标准书号：ISBN 978-7-111-38759-6
定价：35.00 元

电话服务　　　　　　　　　网络服务
客服电话：010-88361066　　机 工 官 网：www.cmpbook.com
　　　　　010-88379833　　机 工 官 博：weibo.com/cmp1952
　　　　　010-68326294　　金 书 网：www.golden-book.com
封底无防伪标均为盗版　　　机工教育服务网：www.cmpedu.com

第 2 版前言

《数控专业英语》第 1 版自 2002 年 8 月出版以来,受到了广大师生的欢迎。结合部分使用原教材师生的意见,在保留原教材风格的前提下,本次再版对原教材进行了较大幅度的修改和调整。

本书由陕西工业职业技术学院王兆奇教授、刘向红副教授担任修订主编,修订框架由王兆奇制订。Unit 1 ~ Unit 4 由北京电子科技职业学院黄桂芸负责修订,其中 Unit 4 重新编写;Unit 5、Unit 6、Unit 25 ~ Unit 30 由陕西工业职业技术学院刘向红负责修订,其中 Unit 5、Unit 6、Unit 29、Unit 30 重新编写;Unit 7 ~ Unit 10 由无锡职业技术学院李晓会负责修订,其中 Unit 7 重新编写;Unit 15 ~ Unit 20 由陕西工业职业技术学院王颖负责修订,其中 Unit 19、Unit 20 重新编写;Unit 11 ~ Unit 14、Unit 21 ~ Unit 24 由王兆奇负责修订,其中 Unit 21 重新编写。王兆奇负责全书修订后的统稿,王颖负责统稿后的资料整理。本书由空军工程大学导弹学院王曙钊教授担任主审。

由于编者水平有限,书中难免还存在错误与不妥之处,希望广大读者批评指正。

编　者

第1版前言

《数控专业英语》是供高等职业技术学院数控、CAD、机电专业学生使用的专业英语教材。通过学习本教材，读者可提高英语阅读水平，掌握常用数控英语词汇，为阅读数控英文资料打下良好基础。

在本教材的编写中，我们精心选编了与数控专业相关的科技信息，涵盖了数控各个发展方向。书中附有插图、生词、短语、注释、练习、译文及阅读材料。全书安排由浅入深，循序渐进。主要内容有：数控的发展历史、数控机床、数控的优点、计算机类型、计算机存储器、输入介质的类型、穿孔带的制作过程、伺服控制系统、MCU及CPU、计算机数字控制、进给与转速、连续路径加工、加工中心种类与组成、刀具监控及在程监测、切削刀具、刀具系统、自适应控制、电火花线切割加工、计算机图形编程、CAD、CAM、CAD/CAM/CNC、柔性制造系统、数控职业范畴、工业机器人入门等。各校可根据教学需要选学其中的相关内容。

本书所载文章全部选自欧美文献原著。陕西工业职业技术学院王兆奇副教授任主编，刘向红任副主编，本书Unit1～Unit13由陕西工业职业技术学院胡梅贻老师编写；Unit4～Unit6由陕西工业职业技术学院段文洁老师编写；Unit7～Unit8由陕西工业职业技术学院夏粉玲老师编写；Unit9～Unit10以及所有阅读材料由常州机械学校汤彩萍老师编写；Unit11～Unit12由华北机电学校虞静老师编写；Unit13～Unit14由北京仪器仪表工业学校黄桂芸老师编写；Unit15由湖南工业职业技术学院周晓宏老师编写；Unit16～Unit23由陕西工业职业技术学院王兆奇教授编写；Unit24～Unit30由陕西工业职业技术学院刘向红老师编写。王兆奇教授对全书进行了总编和修改更正，刘向红负责全书的计算机整理和编辑工作。空军工程大学导弹学院王曙钊教授主审，为本书提出了大量的宝贵意见和建议。编写过程中还得到陕西工业职业技术学院机械系张普礼副教授、数控教研室李善术副教授、赵云龙副教授、杨勇老师、加拿大外籍教师John Whittaker以及澳大利亚外籍教师Christine Story女士的热忱帮助，他们为本书提出了许多建设性的意见和建议，在此一并表示感谢。

由于时间仓促，编者水平所限，疏漏之处在所难免，敬请广大读者及同行批评指正。

编 者
2003年7月于咸阳

CONTENTS

第 2 版前言
第 1 版前言
Unit 1	History of NC	1
Unit 2	Machines Using NC	7
Unit 3	Advantages of NC	14
Unit 4	Modern CNC Machine Tools	20
Unit 5	Elements of CNC Machine Tools	27
Unit 6	CNC Machining Process	37
Unit 7	CNC Programming	44
Unit 8	Servo Controls	49
Unit 9	MCU and CPU	53
Unit 10	CNC	57
Unit 11	Feeds & Speeds	65
Unit 12	Continuous Path	69
Unit 13	Types and Parts of Machining Centers	75
Unit 14	Tool Monitoring and In-Process Gaging	82
Unit 15	Cutting Tools	86
Unit 16	Tooling Systems	91
Unit 17	Adaptive Control	97
Unit 18	Wire-Cut EDM	101
Unit 19	CNC Lathes	107
Unit 20	CNC Milling Machines	113
Unit 21	Spindles of CNC Machine Tools	119
Unit 22	Computer Graphics Programming	125
Unit 23	CAD	129
Unit 24	CAM	133
Unit 25	CAD /CAM/CNC	137
Unit 26	Flexible Machining System	142
Unit 27	Employment Opportunities in NC	147
Unit 28	An Introduction to Industrial Robots	152
Unit 29	Computer Integrated Manufacturing System	155
Unit 30	Manufacturing Technology Facing the 21st Century	161

参考文献 …… 168

Unit 1　　History of NC

Text

Welcome to the world of numerical control (NC). Numerical control has become popular in shops and factories because it helps solve the problem of making manufacturing systems more flexible. In simple terms, a numerical control machine is a machine positioned automatically along a preprogrammed path by means of coded instructions. The key words here are "preprogrammed" and "coded". Someone has to determine what operations the machine is to perform and put that information into a coded form that the NC control unit understands before the machine can do anything. [1] In other words, someone has to program the machine.

Machines may be programmed manually or with the aid of a computer. Manual programming is called manual part programming; programming done by a computer is called computer aided programming (CAP). Sometimes a manual program is entered into the machine's controller via its own keypad. This is known as manual data input (MDI).

Advances in microelectronics and microcomputers have allowed the computer to be used as the control unit on modern numerical control machinery. This computer takes the place of the tape reader found on earlier NC machines. In other words, instead of reading and executing the program directly from punched tape, the program is loaded into and executed from the machine's computer. These machines, known as computer numerical control (CNC) machines, are the NC machines being manufactured today.

In 1947, John Parsons of the Parsons Corporation, began experimenting with the idea of using three-axis curvature data to control machine tool motion for the production of aircraft components. In 1949, Parsons was awarded a U. S. Air Force contract to build what was to become the first numerical control machine. In 1951, the project was assumed by the Massachusetts Institute of Technology (MIT). In 1952, numerical control arrived when MIT demonstrated that simultaneous three-axis movements were possible using a laboratory-built controller and a Cincinnati Hydrotel vertical spindle. [2] By 1955, after further refinements, numerical control became available to industry.

Early NC machines ran off punched cards and tape, with tape becoming the more common medium. Due to time and effort required to change or edit tape, computers were later introduced as aids in programming. Computer involvement came in two forms: computer aided programming languages and direct numerical control (DNC). Computer aided programming languages allowed a part programmer to develop an NC program using a set of universal "pidgin English" commands, which the computer then translated into machine codes and punched into the tape. [3] Direct numerical control involved using a computer as a partial or complete controller of one or more numerical control machines (Fig. 1-1). Although some companies have been reasonably successful at implementing DNC, the expense of computer capability and software and problems associated with coordinating a DNC system renders such systems economically unfeasible for all but the largest companies. [4]

Recently a new type of DNC system called distributive numerical control has been developed (Fig. 1-2). It employs a network of computers to coordinate the operation of a number of CNC machines. Ultimately, it may be possible to coordinate an entire factory in this manner. Distributive numerical control solves some of the problems that exist in coordinating a direct numerical control system. There is another type of distributive numerical control that is a spin-off of the system previously explained. In this system, the NC program is transferred in its entirety from a host computer directly to the machine's controller. Alternately, the program can be transferred from a mainframe host computer to a personal computer (PC) on the shop floor where it will be stored until it is needed. The program will then be transferred from the PC to the machine controller.

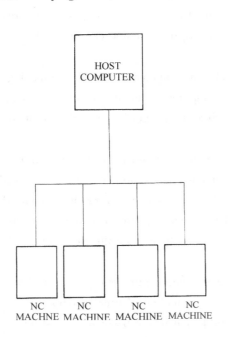

Fig. 1-1 Direct numerical control

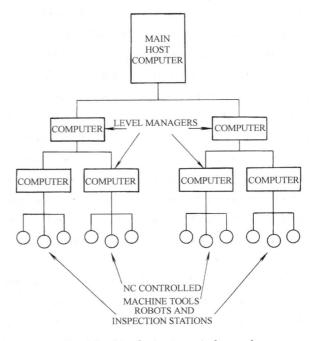

Fig. 1-2 Distributive numerical control

Technical Words

numerical [nju(:)'merikəl]   adj. 数字的
manufacture [ˌmænju'fæktʃə]   v. 制造，加工
                                      n. 制造，制造业；产品
automatically [ˌɔ:tə'mætikəli]   adv. 自动地
program ['prəugræm]   v. （为……）编（制）程序
                                      n. 程序
instruction [in'strʌkʃən]   n. 指令
preprogram [ˌpri:'prəugræm]   v. 预编程序
code [kəud]   v. 编码
                                      n. 代码，编码

information [ˌɪnfəˈmeɪʃən]	n.	信息
manually [ˈmænjuəli]	adv.	手动地，人工地
keypad [ˈkiːpæd]	n.	键区
data [ˈdeɪtə]	n.	数据；参数
input [ˈɪnput]	n.	输入
	v.	输入
microelectronics [ˌmaɪkrəuɪlekˈtrɔnɪks]	n.	微电子学
execute [ˈeksɪkjuːt]	v.	执行
	n.	执行
punch [pʌntʃ]	v.	冲孔，打孔
	n.	冲压机；冲床
load [ləud]	v.	装载，加载
component [kəmˈpəunənt]	n.	部件；零件
technology [tekˈnɔlədʒi]	n.	工艺；技术
spindle [ˈspɪndl]	n.	主轴
medium [ˈmiːdjəm]	n.	方法；媒介
command [kəˈmɑːnd]	n.	命令
	v.	命令，指挥
translate [trænsˈleɪt]	v.	转变为；翻译
controller [kənˈtrəulə]	n.	控制器
software [ˈsɔftwɛə]	n.	软件
coordinate [kəuˈɔːdɪnɪt]	v.	调整；协调
	n.	坐标
network [ˈnetwəːk]	n.	网络
transfer [trænsˈfəː]	v.	传递；改变
	n.	传递，转移
mainframe [ˈmeɪnfreɪm]	n.	主机，大型机
store [stɔː]	v.	存储，贮藏
	n.	储备

Technical Phrases

numerical control (NC)	数字控制（数控）
control unit	控制装置，控制单元
manual programming	手工编程
computer aided programming (CAP)	计算机辅助编程
manual data input (MDI)	手动数据输入
tape reader	读带机
punched tape	穿孔带
computer numerical control (CNC)	计算机数字控制

machine tool	机床
punched card	穿孔卡
direct numerical control (DNC)	直接数字控制
distributive numerical control (DNC)	分布式数字控制
host computer	主机
personal computer (PC)	个人计算机

Notes

(1) Someone has to **determine** what operations *the machine is to perform* and **put** that information into a coded form *that the NC control unit understands* before the machine can do anything.

主句中 Someone 作主语，has to determine 与（has to） put 是两个并列的谓语。定语从句 *the machine is to perform* 修饰 operations；定语从句 *that the NC control unit understands* 修饰 form。

(2) In 1952, numerical control arrived when MIT demonstrated that simultaneous three-axis movements were possible using a laboratory-built controller and a Cincinnati Hydrotel vertical spindle.

现在分词短语 using a laboratory-built controller and a Cincinnati Hydrotel vertical spindle 在句中作状语。

(3) Computer aided programming languages allowed a part programmer to develop an NC program using a set of universal "pidgin English" commands, *which the computer then translated into machine codes and punched into the tape.*

using a set of universal "pidgin English" commands 是现在分词短语作状语；which 引导的非限定性定语从句修饰 NC program。

(4) Although some companies have been reasonably successful at implementing DNC, the expense of computer capability and software and problems associated with coordinating a DNC system renders such systems economically unfeasible for all but the largest companies.

主句的主语较长，其中心词是 expense；computer capability，software，problems associated with coordinating a DNC system 作 expense 的定语；谓语动词是 render。此外，associated with coordinating a DNC system 是过去分词短语，作 problems 的后置定语；but 为介词，意为"除……之外"。

Exercises

(1) Place a "T" after sentences that are true and an "F" after those that are false.

1) An NC machine is positioned automatically along a preprogrammed path by means of coded instructions.

2) CNC stands for *computer numerical control*.

3) According to the last sentence of paragraph 5, DNC systems are economically unfeasible for all companies.

4) In a distributive numerical control system, it is impossible for the NC program to be trans-

ferred in its entirety from a host computer directly to the machine's controller.

(2) Fill in the blanks according to the text with the words given below. Make changes if necessary.

   manually punch controller network

1) Machines may be programmed _____ or with the aid of a computer.

2) A manual program is sometimes entered into the machine's _____ via its own keypad.

3) Early NC machines could run off _____ cards and tape, with tape being the more common medium.

4) A distributive numerical control system uses a _____ of computers to coordinate the operation of many CNC machines.

【参考译文】

第1课　数控的发展历史

　　欢迎来到数字控制（NC）世界。由于数字控制使制造系统柔性更强，因而已广泛应用于工厂与车间。简而言之，数控机床是通过代码指令使机床沿着预编程轨迹自动定位的机床，这里的关键词是"预编程"与"代码化"。机床运行前，必须有人确定需要机床执行什么操作，并将此信息译成数控装置能够识别的代码。换句话说，必须有人对机床编程。

　　人们可以手工编写加工程序，称其为手工编程；亦可借助计算机对机床编程，称为计算机辅助编程（CAP）。手工编好的程序有时可从机床键盘送入控制器，这叫做手动数据输入（MDI）。

　　微电子技术与微型计算机的发展已使计算机可用于现代数控机床的控制单元，取代了早期 NC 机床的读带机。换句话说，程序是由机床上的计算机存取和执行的，而不再直接取自穿孔带。这种计算机数控（CNC）机床就是当今制造的数控（NC）机床。

　　1947 年，Parsons 公司的 John Parsons 着手进行一项试验，他想用三轴曲度数据来操纵机床加工飞机零件。1949 年，Parsons 公司与美国空军签订了制造第一台数控机床的合同。1951 年，美国麻省理工学院承担了这一项目。1952 年，麻省理工学院（MIT）使用实验室制造的控制器和辛辛那提立式主轴展示三轴联动获得成功，这标志着数控时代的到来。到了 1955 年，几经改进之后，数控技术开始应用于工业生产。

　　早期的 NC 机床能运行穿孔卡与穿孔带，二者中以穿孔带更为通用。但是，鉴于更换、编辑纸带费时费力，后来便采用计算机作为编程的辅助工具。计算机在数控中的应用有两种形式：一是计算机辅助编程语言，二是实施直接数字控制（DNC）。有了计算机辅助编程语言，程序员可用一套通用"混杂英语"命令编写 NC 程序，然后由计算机将其释译为机器码并制成穿孔带。直接数字控制是指用一台计算机对一台或多台数控机床实施部分或整体控制（见图 1-1）。虽然有些公司运用 DNC 已获得成功，但是，扩大计算机容量、购买软件、协调 DNC 系统等花费使这种系统并不适合所有公司，而只适用于一些大公司。

　　最近，一种叫做分布式数字控制的新型 DNC 系统（见图 1-2）已经开发出来，它用计算机网络来协调多台 CNC 机床的运行。这种方式最终有可能用来协调整个工厂的运转。这种分布式数字控制方法解决了协调直接数字控制系统时遇到的一些难题。在此基础上，人们还开发出另一种分布式数字控制系统。在这个系统中，整个 NC 程序可从主机直接传输到机床控制器。另外，该系统也可在必要时将程序从主机传输到车间的个人计算机（PC）上储存起来，以便需要时再传输到机床控制器。

Unit 2   Machines Using NC

Text

Early machine tools were designed so that the operator was standing in front of the machine while operating the controls. This design is no longer necessary, since in NC the operator no longer controls the machine tool movements. On conventional machine tools, only about 20 percent of the time was spent removing material. With the addition of electronic controls, actual time spent removing metal has increased to 80 percent and even higher. It has also reduced the amount of time required to bring the cutting tool into each machining position.

In the past, machine tools were kept as simple as possible in order to keep their costs down. Because of the ever-rising cost of labor, better machine tools, complete with electronic controls, were developed so that industry could produce more and better goods at prices which were competitive with those of offshore industries. [1]

NC is being used on all types of machine tools, from the simplest to the most complex. The most common machine tools are the single-spindle drilling machine, engine lathe, milling machine, turning center, and machining center.

1. Single-Spindle Drilling Machine

One of the simplest numerically controlled machine tools is the single-spindle drilling machine (Fig. 2-1). Most drilling machines are programmed on three axes:

a. The X-axis controls the table movement to the right and left.

b. The Y-axis controls the table movement toward or away from the column.

c. The Z-axis controls the up or down movement of the spindle to drill holes to depth.

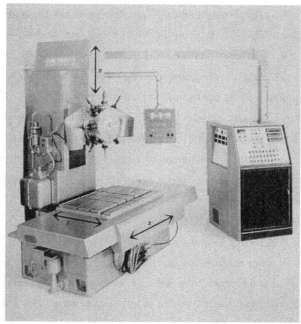

Fig. 2-1   Numerically controlled single-spindle drilling machine

2. Engine Lathe

The engine lathe, one of the most productive machine tools, has always been a very efficient means of producing round parts (Fig. 2-2). Most lathes are programmed on two axes:

a. The X-axis controls the cross motion (in or out) of the cutting tool.

b. The Z-axis controls the carriage travel toward or away from the headstock.

3. Milling Machine

The milling machine has always been one of the most versatile machine tools used in industry (Fig. 2-3). Operations such as milling, contouring, gear cutting, drilling, boring, and reaming are only a few of the many operations which can be performed on a milling machine. The milling machine can be programmed on three axes:

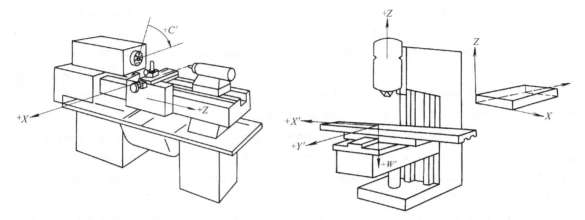

Fig. 2-2 The engine lathe cutting tool moves only on the X and Z axes

Fig. 2-3 The vertical knee and column milling machine

a. The X-axis controls the table movement left or right.
b. The Y-axis controls the table movement toward or away from the column.
c. The Z-axis controls the vertical (up or down) movement of the knee or spindle.

4. Turning Center

Turning Centers were developed in the mid-1960s after studies showed that about 40 percent of all metal cutting operations were performed on lathes. These numerically controlled machines are capable of greater accuracy and higher production rates than were possible on the engine lathe.

The basic turning center operates on only two axes:
a. The X-axis controls the cross motion of the turret head.
b. The Z-axis controls the lengthwise travel (toward or away from the headstock) of the turret head.

5. Machining Center

Machining centers (Fig. 2-4) were developed in the 1960s so that a part did not have to be moved from machine to machine in order to perform various operations. These machines greatly increased production rates because more operations could be performed on a workpiece in one setup. There are two main types of machining centers, the horizontal and the vertical spindle types.

a. The horizontal spindle machining center (Fig. 2-5) operates on three axes:
a) The X-axis controls the table movement left or right.
b) The Y-axis controls the vertical movement (up or down) of the spindle.
c) The Z-axis controls the horizontal movement (in or out) of the spindle.

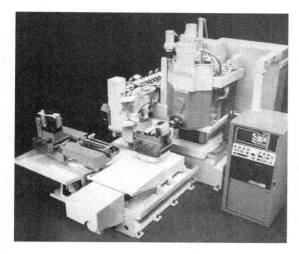

Fig. 2-4　Machining center

Fig. 2-5　Horizontal spindle machining center

b. The vertical spindle machining center (Fig. 2-6) operates on three axes:
a) The X-axis controls the table movement left or right.
b) The Y-axis controls the table movement toward or away from the column.
c) The Z-axis controls the vertical movement (up or down) of the spindle.

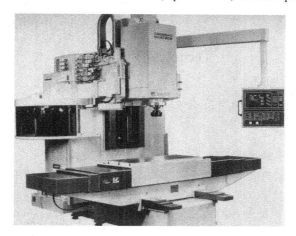

Fig. 2-6　Vertical spindle machining center

Technical Words

operator [ˈɔpəreitə]	n.	操作者
operate [ˈɔpəreit]	v.	操作；运行
electronic [ˌilekˈtrɔnik]	adj.	电子的
drill [dril]	v.	钻孔
	n.	钻孔机
lathe [leið]	n.	车床

	v.	用车床加工
milling ['miliŋ]	n.	铣削，铣加工
turning ['tə:niŋ]	n.	车削
table ['teibl]	n.	工作台
column ['kɔləm]	n.	立柱
part [pɑ:t]	n.	零件，工件
carriage ['kæridʒ]	n.	（机床的）滑板；刀架
headstock ['hedstɔk]	n.	主轴箱
contouring [kən'tuəriŋ]	n.	成形加工
bore [bɔ:]	v.	镗（穿、扩、钻）孔
boring ['bɔ:riŋ]	n.	镗孔；镗削加工
ream [ri:m]	v.	铰孔
reaming ['ri:miŋ]	n.	铰孔
knee [ni:]	n.	升降台
accuracy ['ækjurəsi]	n.	精确性，准确度，精度
workpiece ['wə:kpi:s]	n.	工件
setup [setʌp]	n.	安装；设备；机构
horizontal [ˌhɔri'zɔntl]	adj.	水平的
vertical ['və:tikəl]	adj.	垂直的，直立的

Technical Phrases

cutting tool	刀具
drilling machine	钻床
turning center	车削中心
machining center	加工中心
engine lathe	卧式车床（普通车床）
cross motion	横向运动
gear cutting	齿轮加工
metal cutting	金属切削
production rate	生产率
turret head	转塔头
lengthwise travel	纵向运动

Note

Because of the ever-rising cost of labor, better machine tools, complete with electronic controls, were developed so that industry could produce more and better goods at prices *which were competitive with those of offshore industries*.

句子主语是 better machine tools，谓语是 were developed；so that 引导结果状语从句；

which 引导的定语从句修饰 goods。

Exercises

(1) Fill in the blanks according to the text with the words given below. Make changes if necessary.

  electronic axis cost operation program material

1) On conventional machine tools, only 20% of the time is spent removing _____.

2) Due to the addition of _____ controls, actual time spent removing metal has increased to 80% and even higher.

3) Machine tools used to be kept as simple as possible in order to keep their _____ down.

4) Most drilling machines can be _____ on three axes.

5) Milling, contouring, drilling, boring, and reaming are just a few of the _____ performed on a milling machine.

6) A basic turning center operates on only two _____.

(2) Explanation：

  NC CNC CAP MDI MIT DNC PC

【参考译文】

第 2 课　数 控 机 床

老式机床是按操作工站立在机床前进行操作来设计的。现在不需要这种设计了，因为操作工并不直接操纵数控机床的运行。在传统机床上加工时，只有约 20% 的时间用于切削材料。随着电控设备的加入，实际切削材料的时间已增至 80% 以上，并且缩短了将刀具送入每个加工位置的时间。

过去，人们尽量使机床结构简单，以便降低成本。由于劳动成本日益上涨，人们研制出性能更好的机床，并配有电控设备，这样企业可以生产更多更好的价格较低的产品，以和国际上的产品相竞争。

从最简单到最复杂的机床都会用到数控技术。最常见的机床有：单轴钻床、卧式车床、铣床、车削中心及加工中心。

1. 单轴钻床

单轴钻床是最简单的数控机床之一（见图 2-1）。多数数控钻床可在三个坐标轴上编程：

1) X 轴控制工作台左右运动。

2) Y 轴控制工作台靠近或离开立柱。

3) Z 轴控制主轴上下运动，确定孔的加工深度。

2. 卧式车床

卧式车床是生产效率最高的机床之一，它是加工回转体零件时非常有效的工具。大部分数控车床可在两个坐标轴上编程（见图 2-2）。

1) X 轴控制刀具横向运动（切入、切出）。
2) Z 轴控制机床拖板靠近或离开主轴箱。

3. 铣床

铣床是工业生产中加工方式最多的机床之一（见图 2-3），像铣削、成形加工、齿轮加工、钻、镗、铰等只是可在铣床上进行的一少部分加工。数控铣床可在三个坐标轴上编程：

1) X 轴控制工作台左右运动。
2) Y 轴控制工作台靠近或离开立柱。
3) Z 轴控制升降台或主轴垂直（上下）运动。

4. 车削中心

研究表明，整个金属材料的切削操作大约 40% 是在车床上进行的。早在 20 世纪 60 年代中期，人们就开始研制车削中心。这种数控机床具有比普通车床更高的加工精度和生产效率。

一般数控车削中心仅在两轴上工作：

1) X 轴控制转塔头横向运动。
2) Z 轴控制转塔头纵向运动（靠近或离开主轴箱）。

5. 加工中心

加工中心（见图 2-4）是 20 世纪 60 年代发展起来的。有了加工中心，人们不必把零件从一台机床转移到另一台机床就能完成各种加工。由于工件经过一次装夹后便能进行多种加工，所以大大提高了生产效率。加工中心主要有两类：卧式加工中心与立式加工中心。

（1）卧式加工中心（见图 2-5）可在三个坐标轴上工作：

1) X 轴控制工作台左右运动。
2) Y 轴控制主轴垂直（上下）运动。
3) Z 轴控制主轴水平运动（切入、切出）。

（2）立式加工中心（见图 2-6）在三个坐标轴上工作：

1) X 轴控制工作台左右运动。
2) Y 轴控制工作台靠近或离开立柱。
3) Z 轴控制主轴垂直（上下）运动。

【Reading Material】

Work Coordinates System

These machines have three linear axes named X, Y, and Z. The X-axis moves the table left and right, the Y-axis moves it to and from the operator and the Z moves the milling head up and down. The machine zero position is the upper right corner of the mill table. All moves from this point are in a negative machine direction. If a rotary table is connected, an additional A-axis work offset is provided.

The work offset display is found on the offset display by pushing the PAGE UP key. You can display and manually enter work offsets from here. The work coordinate systems on a control with a fifth axis have all been expanded to accommodate B, the fifth axis. Work coordinate offsets can be set for the B axis in the offset display. Note that the auxiliary axes C, U, V, and W do not have any offsets; they are always programmed in machine coordinates.

The Home or Machine-Zero position is X_0, Y_0, and Z_0. Travel of these axes is limited in the negative direction by stored stroke limits defined in the parameters. Travel in the positive direction for the X and Y axes is limited simply to values less than zero. Positive travel for the Z-axis is limited to the highest position used for tool changing (about Z4.5). In addition, positive travel for all axes is limited by the home switch which acts as a limit switch.

Before a tool can machine your part, the control must know where your part is. The work coordinate system tells the control the distance from the work zero point of your part to the machine zero position. The work zero point of the part is decided by the programmer and usually is the common point which all print dimensions are referenced from. The machine zero position is fixed by the machine on power up and does not change. The operator must determine this distance and enter the value.

This control automatically chooses the G54 system on power up. If you do not wish to use this system, zero out the values in the G54 X, Y, and Z or select another work offset.

The G54 through G59 or G110 through G129 offsets can be set by using the PART ZERO SET key. Position the axes to the work zero point of your part. Using the cursor, select the proper axis and work number. Press the PART ZERO SET key and the current machine position will be automatically stored in that address. This will work with only the work zero offset display selected. Note that entering a nonzero Z work offset will interfere with the operation of an automatically entered tool length offset.

Unit 3 Advantages of NC

Text

Recent studies show that of the amount of time an average part spent in a shop, only a fraction of that time was actually spent in the machining process. Let us assume that a part spent 50 hours from the time it arrived at a plant as a rough casting or bar stock to the time it was a finished product.[1] During this time, it would be on the machine for only 2.5 hours and be cut for only 0.75 hour. The rest of the time would be spent on waiting, moving, setting up, loading, unloading, inspecting the part, setting speeds and feeds, and changing cutting tools.

NC reduces the amount of non-chip-producing time by selecting speeds and feeds, making rapid moves between surfaces to be cut, using automatic fixtures, automatic tool changing, controlling the coolant, in-process gaging, and loading and unloading the part.[2] These factors, plus the fact that it is no longer necessary to train machine operators, have resulted in considerable savings throughout the entire manufacturing process and caused tremendous growth in the use of NC.[3] Some of the major advantages of NC are as follows:

(1) There is automatic or semiautomatic operation of machine tools. The degree of automation can be selected as required.

(2) Flexible manufacturing of parts is much easier. Only the tape needs changing to produce another part.

(3) Storage space is reduced. Simple workholding fixtures are generally used, reducing the number of jigs or fixtures which must be stored.

(4) Small part lots can be run economically. Often a single part can be produced more quickly and better by NC.

(5) Nonproductive time is reduced. More time is spent on machining the part, and less time is spent on moving and waiting.

(6) Tooling costs are reduced. In most cases complex jigs and fixtures are not required.

(7) Inspection and assembly costs are lower. The quality of the product is improved, reducing the need for inspection and ensuring that parts fit as required.

(8) Machine utilization time is increased. There is less time that a machine tool is idle because workpiece and tool changes are rapid and automatic.

(9) Complex forms can easily be machined. The new control unit features and programming capabilities make the machining of contours and complex forms very easy.

(10) Parts inventory is reduced. Parts can be made as required from the information on the punched tape.

Since the first industrial revolution, about 200 years ago, NC has had a significant effect on the industrial world. The developments in the computer and NC have extended a person's mind and muscle. The NC unit takes symbolic input and changes it to useful output, expanding a person's concepts into creative and productive results. NC technology has made such rapid advances that it is being used in almost every area of manufacturing, such as machining, welding, pressworking, and assembly.

If industry's planning and logic are good, the second industrial revolution will have as much or more effect on society as the first industrial revolution had. As we progress through the various stages of NC, it is the entire manufacturing process which must be kept in mind.

Computer-assisted manufacturing (CAM) and computer-integrated machining (CIM) are certainly where the future of manufacturing lies, and considering the developments of the past, it will not be too far in the future before the automated factory is a reality.[4]

Developed originally for use in aerospace industries, NC is enjoying widespread acceptance in manufacturing. The use of CNC machines continues to increase, becoming visible in most metalworking and manufacturing industries. Aerospace, defense contract, automotive, electronic, appliance, and tooling industries all employ numerical control machinery. Advances in microelectronics have lowered the cost of acquiring CNC equipment. It is not unusual to find CNC machinery in contract tool, die, and moldmaking shops. With the advent of low cost OEM (original equipment manufacturer) and retrofit CNC vertical milling machines, even shops specializing in one-of-a-kind prototype work are using CNCs.

Although numerical control machines traditionally have been machine tools, bending, forming, stamping, and inspection machines have also been produced as numerical control systems.

Technical Words

load [ləud]	v.	装载
inspect [in'spekt]	v.	检查
feed [fiːd]	n.	进给
chip [tʃip]	n.	碎片，碎屑
	v.	切成碎片
fixture ['fikstʃə]	n.	夹具
coolant ['kuːlənt]	n.	切削液
process ['prəuses]	n.	过程，步骤
	v.	加工，处理
gage [geidʒ]	v.	检验，校准
semiautomatic ['semiˌɔːtə'mætik]	adj.	半自动的
automation [ˌɔːtə'meiʃən]	n.	自动化
storage ['stɔridʒ]	n.	贮藏，存储
workholding ['wəːkhəuldiŋ]	n.	工件夹紧
jig [dʒig]	n.	夹具
assembly [ə'sembli]	n.	装配，组装

contour ['kɔntuə]	n. 轮廓
output ['autput]	n. 输出
welding ['weldiŋ]	n. 焊接
logic ['lɔdʒik]	n. 逻辑；逻辑性
integrate ['intigreit]	v. 使一体化，集成
metalworking ['metəl‚wə:kiŋ]	n. 金属加工术
aerospace ['ɛərəuspeis]	n. 航空航天技术
automotive [‚ɔ:tə'məutiv]	adj. 汽车的
tooling ['tu:liŋ]	n. 加工；刀具，工具
equipment [i'kwipmənt]	n. 装置，设备，装备
die [dai]	n. 模具，冲模
retrofit ['retrəfit]	n. 改型（装；进）；（式样）翻新
bending ['bendiŋ]	n. 弯曲
forming ['fɔ:miŋ]	n. （成形）加工
stamping ['stæmpiŋ]	n. 冲压

Technical Phrases

machining process	加工过程
rough casting	铸造毛坯
bar stock	棒料
finished product	成品
non-chip-producing time	非切削时间
tool changing (tool change)	换刀
in-processing gaging	在程检测
tooling cost	刀具加工成本
computer-assisted manufacturing (CAM)	计算机辅助制造
computer-integrated machining (CIM)	计算机集成制造
tooling industry	刀具业
prototype work	标准工件
mold making	模具制作

Notes

(1) Let us assume that a part spent 50 hours from the **time** *it arrived at a plant as a rough casting or bar stock* to the **time** *it was a finished product*.

句中 that 引导的是宾语从句；介词短语 from the time... to the time... 中，time 后分别是省略 that 的定语从句。

(2) NC reduces the amount of non-chip-producing time by selecting speeds and feeds, making rapid moves between surfaces to be cut, using automatic fixtures, automatic tool changing, control-

ling the coolant, in-process gaging, and loading and unloading the part.

此句虽长，但结构简单：NC reduces...by...。by 表示方法与手段。

(3) These factors, plus the fact that it is no longer necessary to train machine operators, **have resulted in** considerable savings throughout the entire manufacturing process and **caused** tremendous growth in the use of NC.

句子主语是 These factors，谓语有两个：have resulted in 与（have）caused；the fact 后由 that 引出同位语从句。

(4) Computer-assisted manufacturing (CAM) and computer-integrated machining (CIM) are certainly where the future of the manufacturing lies, and considering the developments of the past, it will not be too far in the future before the automated factory is a reality.

现在分词短语 considering the developments of the past 作状语。

Exercises

(1) Place a "T" after sentences that are true and an "F" after those that are false.

1) According to the first paragraph, of the amount of time a part spent in a shop, only a fraction of that time was actually spent in the machining process.

2) Productive time is reduced. Less time is spent on machining the part, and more time is spent on moving and waiting.

3) If industry's planning and logic are good, the second industrial revolution will have as much effect on society as the first industrial revolution had.

4) It is usual to find CNC machinery in contract tool, die, and mold making shops, even shops specializing in one-of-a-kind prototype work are using CNCs.

(2) Fill in the blanks according to the text with the words given below. Make changes if necessary:

significant   stamping   rough casting   utilization   finish   non-chip
automatic   fixture

1) A part spent 50 hours from the time it arrived at a plant as a _____ or bar stock to the time it was a (an) _____ product.

2) Machine _____ time is increased because workpiece and tool changes are rapid and _____.

3) Since the first industrial revolution, NC has had a _____ effect on the industrial world.

4) Although NC machines traditionally have been machine tools, bending, forming, _____, and inspection machines have also been produced as NC systems.

5) NC reduces the amount of _____ producing time by selecting speeds, feeds, using automatic _____ and many other ways.

(3) Answer the following questions:

1) How would the time be spent on a part from the time it arrived at a plant as a rough casting to the time it was a finished product?

2) What are the major advantages of NC?

3) Which industries did NC develop originally for use in? Which industries employ numerical control machinery now?

【参考译文】

第3课 数控技术的优点

近期研究表明，在加工零件时，只有一少部分时间真正用于加工工件。假定把毛坯或棒料加工为成品零件需要50小时，那么，工件在机床上的时间只有2.5小时，用于切削的时间仅有0.75小时，其余时间都用在等待加工、运输、安装、装夹、拆卸、检验、设置转速与进给、换刀上。

通过转速与进给量的设定、刀具在待切削表面间的快速移动、自动夹具的运用、刀具的自动切换、切削液的控制、在线检测及零件的自动装卸，NC缩短了非切削加工时间。上述因素，加之无需培训机床操作工，使整个加工过程耗时大量减少，促进了数控的推广应用。NC的主要优点如下：

1) 机床可以自动运行，亦可半自动运行，使用时按需要选取。
2) 零件的柔性制造更为简便。更换纸带即可加工不同零件。
3) 减小存储空间。通常采用简单的工装夹具，因而减少了存储夹具的数量。
4) 加工小批量零件非常经济。用NC机床加工出的单件质量好且加工快。
5) 缩短非加工时间。花在加工零件上的时间多了，用于运输和等待的时间短了。
6) 降低加工成本。数控机床在多数情况下无需复杂夹具。
7) 降低检验与装配成本。产品质量提高了，因而减少了检验的必要，保证零件能按需装配。
8) 增加机床使用时间。由于能迅速自动地更换工件与刀具，机床闲置时间缩短了。
9) 便于加工复杂形状的工件。新型控制器功能多、编程能力强，因而轮廓加工与复杂形状的加工非常方便。
10) 减少零件存货。人们能根据穿孔带上的信息加工所要求的零件。

自从约200年前的第一次工业革命以来，NC已对工业界产生了深远的影响。计算机的发展与NC技术的应用拓展了人类的思维与体能。NC装置将符号式输入转换为有用输出，将人们的想法创造性地转换为生产结果。NC技术已经取得了迅速发展，它几乎应用于制造业的每一个领域，例如机加工、焊接、压力加工和装配。

如果工业规划合理、逻辑正确，与第一次工业革命一样，第二次工业革命将对人类社会产生同样甚至更大的影响。在我们从事NC不同阶段的工作时，必须牢记NC技术贯穿于整个加工过程的始终。

回顾数控技术的发展历史，我们相信，计算机辅助制造（CAM）与计算机集成制造（CIM）是制造业的发展趋势，自动化工厂在不远的未来将成为现实。

NC虽然起源于航空工业，但它在当今制造业中得到了广泛应用。很明显，在大多数金属加工与制造业中，CNC机床的应用持续增长，航空、国防、汽车、电子、仪器、机床行业都采用数控机床。微电子技术的发展降低了 CNC 装置的成本,在工具、模具与制模车间使用 CNC 机床已是一件常事。随着低成本的 OEM(设备制造商)与立式 CNC 铣床得到改进,甚至那些专门生产某种标准工件的车间也会使用 CNC 机床。

虽然数控机床传统上是指金属切削机床,但折弯机、成形机、冲压机和测量机也使用数字控制系统。

【Reading Material】

Basic Programming and Operation Principles

In an "NC" (Numerically Controlled) machine, the tool is controlled by a code system that enables it to be operated with minimal supervision and with a great deal of repeatability. "CNC" (Computerized Numerical Control) is the same type of operating system, with the exception that the machine tool is monitored by a computer.

The same principles used in operating a manual machine are used in programming an NC or CNC machine. The main difference is that instead of cranking handles to position a slide to a certain point, the dimension is stored in the memory of the machine control once. The control will then move the machine to these positions each time the program is run.

The operation of the VF-Series Vertical Machining Center requires that a part program be designed, written, and entered into the memory of the control. The most common way of writing part programs is off-line, that is, away from the CNC in a facility that can save the program and send it to the CNC control. The most common way of sending a part program to the CNC is via an RS-232 interface. The HAAS VF-Series Vertical Machining Center has an RS-232 interface that is compatible with most existing computers and CNCs.

In order to operate and program a CNC controlled machine, a basic understanding of machining practices and a working knowledge of math is necessary. It is also important to become familiar with the control console and the placement of the keys, switches, displays, etc., which are pertinent to the operation of the machine.

Unit 4 Modern CNC Machine Tools

Text

From the clays of the first NC milling machine,[1] there have been many applications for NC technology, ranging from milling, turning, and electric discharge machining (EDM) to laser, flame and plasma cutting, punching and nibbling, forming, bending, grinding, inspection, and robotics.

Although aerospace is still one of the principal industries that require and use NC technology extensively, other industries have also embraced it. [2] Because of the continuing advances in computers and their affordability, the cost of NC technology has been dropping rapidly. Now, even small machine shops and small specialty industries have come to acquire this technology.

Today you can find NC products in many areas ranging from metal-working and automotive to electronics, appliances, engraving, sign making, jewelry design, and furniture manufacturing.

Turning (See Fig. 4-1)

Turning utilizes a cutter that moves perpendicularly through the center plane of a rotating work-piece. The part shape depends on the shape of the tool and the operations performed to obtain the finished part.

Milling

The process of milling involves the use of a rotating cutter to remove material from a work-piece. Single-or multiple-axis control moves can generate either simple two-dimensional patterns or profiles, or complex three-dimensional shapes.

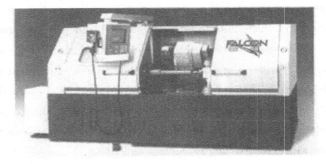

Fig. 4-1 A Cincinnati Milacron Falcon Turning Center

Wire-Cut EDM

Electrical discharge machining, or wire-cut EDM, uses an electrical discharge from a thin wire to achieve fine cuts through hard metal parts. Most EDM machines use two parallel planes in which each cutting point can move independently of the other. [3] This is useful in producing tapered pieces used in the production of punch dies for stamping. Such specialty operations are commonplace in today's high-volume machining environments.

Machining Center (See Fig. 4-2 and Fig. 4-3)

The flexibility and versatility of nu-

Fig. 4-2 A Cincinnati Milacron Universal 5-Axis Machining Center with automatic pallet changer and chip conveyor option

merical control have led to the development of a new type of machine tool called the machining center. Using simpler work holding fixtures and fewer cutting tools, this machine does the work formerly done on several machines.

The machining center is provided with an automatic tool changer. On a command from the NC system, the tool change arm located above the spindle rotates clockwise, simultaneously gripping the tool in the spindle and another tool in an interchange station located on the face of the machine which is used to store the tool temporarily.[4] The arm then moves forward, removing the tools from the spindle and from the interchange station.

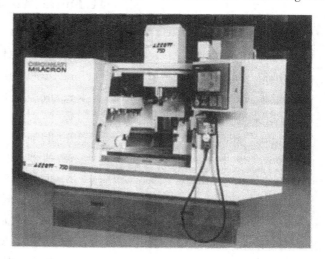

Fig. 4-3　A Cincinnati Milacron Arrow 750 Machining Center with an A-Axis rotary option

After rotating clockwise 180 degrees, the arm retracts, inserts the new tool in the spindle, and places the used tool in the interchange system, the arm then returns to the original position.[5] The tool change operation may be completed in 5 seconds. The tool drum can hold a large number of different tools. Each tool holder being coded, the tools can be selected in a random order and in a sequence.

Machining centers can also be available without NC, however, provided with numerical control the potential of these machines can be fully realized.

Coordinate Measuring Machines

Coordinate measuring machines (CMM) were developed because of the need for higher quality and increased productivity. The inspection techniques that catch defects only after parts have been made are no longer acceptable.[6] This shift from detection to prevention is taking inspection to the manufacturing floor to provide timely feedback for process control and correction. CMMs are designed to check the accuracy of the part as it is being manufactured, based on the information about the part programmed into its control unit.

Laser, Flame, and Plasma Cutting

Laser, flame, and plasma cutting use a powerful beam of light, a concentrated flame, or a plasma arc, respectively, to remove material. Depending on the target and thickness of the material, each application has certain advantages.

Punching and Nibbling

Punching and nibbling are used to cut patterns in sheets of metal by the use of punch dies. Repeated punches along a path achieve a nibbling effect that allows cutting of complex patterns, which would otherwise be very difficult with conventional means.[7] Forming is also typical applications of CNC punch machines.

Technical Words

laser[ˈleizə]	n.	激光
plasma[ˈplæzmə]	n.	等离子体
nibbling [ˈnibliŋ]	n.	步冲轮廓机
forming[fɔːmiŋ]	n.	成型加工
bend[bend]	v.	使弯曲
grinding[ˈgraindiŋ]	n.	磨削
robotics[rəuˈbɔtiks]	n.	机器人；机器人学
cutter[ˈkʌtə]	n.	刀具
rotate[ˈrəuteit]	v.	旋转，转动
axis[ˈæksis]	n.	轴；坐标轴
variation[ˌvɛəriˈeiʃn]	n.	变更，变化，变异
die[dai]	n.	硬模，钢型，冲模
feedback[ˈfiːdbæk]	n.	反馈

Technical Phrases

milling machine	铣床
electric discharge machining (EDM)	电火花加工机床
machining center	加工中心
cutting tool	切削工具
automatic tool changer	自动刀库
tool drum	鼓轮式刀库
coordinate measuring machines	坐标测量机
flame and plasma cutting	火焰切割和等离子切割
punch die	冲模
punch machine	打孔机
nibbling machine	步冲轮廓机
wire-cut EDM	线切割电火花加工机床

Notes

(1) From the clays of the first NC milling machine, …
自从第一台 NC 铣床诞生以来，……

(2) Although aerospace is still one of the principal industries that require and use NC technology extensively, other industries have also embraced it.
句中的 that require and use NC technology extensively 为定语从句，修饰 the principal industries；one of the principal industries 可译为 "主要工业领域之一"。embrace 原意是 "包含"，这里引申为 "使用"。

(3) Most EDM machines use two parallel planes in which each cutting point can move inde-

pendently of the other.

句中 in which each cutting point can move independently of the other 是定语从句，修饰 two parallel planes；two parallel planes 可译为"两个平行的加工面"。

（4）On a command from the NC system, the tool change arm located above the spindle rotates clockwise, simultaneously gripping the tool in the spindle and another tool in an interchange station located on the face of the machine which is used to store the tool temporarily.

主句的主干成分是 the tool change arm...rotates clockwise。此外，which is used to store the tool temporarily 为定语从句，修饰 an interchange station（换刀位置）。

（5）After rotating clockwise 180 degrees, the arm **retracts**, **inserts** the new tool in the spindle, and **places** the used tool in the interchange system, the arm then **returns** to the original position.

句中 the arm 为主语，四个谓语动词分别是 retracts, inserts, places, returns；句中 used tool 可译为"旧刀具"。

（6）The inspection techniques that catch defects only after parts have been made are no longer acceptable.

The inspection techniques...are no longer acceptable. 是句子的主干成分，而 that catch defects... 为定语从句修饰 techniques；no longer 为"不再"之意。

（7）Repeated punches along a path achieve a nibbling effect that allows cutting of complex patterns, which would otherwise be very difficult with conventional means.

句中 that 引导的定语从句修饰 a nibbling effect，which 引导的非限定性定语从句修饰前面整个句子；conventional means 可译为"传统的加工方法"。

Exercises

（1）Fill in the blanks according to the text with the words given below. Make changes if necessary.

versatility   accuracy   three-dimensional   CMM   flexibility

1）Single-or multiple-axis control moves can generate either simple two-dimensional patterns or profiles, or complex _____ shapes.

2）The _____ and _____ of numerical control have led to the development of a new type of machine tool called the machining center.

3）_____ are designed to check the _____ of the part as it is being manufactured, based on the information about the part programmed into its control unit.

（2）Answer the following questions:

1）What shapes can be generated by means of single-or multiple-axis control moves of milling?

2）How to achieve fine cuts through hard metal parts by wire EDM?

3）How to remove material using laser, flame, and plasma cutting?

【参考译文】

第 4 课　现代数控机床

自从第一台 NC 铣床诞生以来，数控技术一直得到广泛应用，其范围从铣削、车削、电火花加工到激光加工、火焰及等离子切割、冲压与步冲轮廓加工、成型、弯曲、磨削、检验和机器人等领域。

尽管航空工业仍然是 NC 技术需求与应用最为广泛的主要工业领域之一，但是 NC 技术在其他工业领域也有所运用。由于计算机性能和人们购买力的不断提高，NC 技术的成本也随之快速下降，一些小型制造厂及小型特种工业目前也开始应用这项技术。

如今 NC 技术应用很广，其产品涉及金属加工、汽车行业、电子产品、器具、雕刻、标牌制作、珠宝设计及家具制造等。

车削加工（见图 4-1）

车削加工是使车刀垂直通过旋转工件中心面进行的加工。成品工件的形状取决于刀具形状以及采用何种加工操作。

铣削加工

铣削是指使用旋转刀具切削工件的加工方法。单轴或多轴控制可加工出简单二维图形、轮廓图形或者复杂的三维形状。

线切割机床

电火花加工也叫线切割加工，是用金属丝放电实现高硬度零件的精加工方法。大多数线切割机床有两个平行加工面，两个加工点在加工过程中的移动相互独立，这在加工用于冲压模压印过程的锥形件时非常有用。这些专门化操作在当今大批量生产环境下是很常见的。

加工中心（见图 4-2 和图 4-3）

利用数控技术的灵活性好、用途广泛，人们又研制出一种新型机床，这就是加工中心。加工中心能够使用更简单的夹具和更少的刀具即可完成以前需要数台机床才能完成的加工任务。

加工中心配有自动换刀装置。一旦收到数控系统发来的指令，主轴上方的换刀机械手顺时针旋转，一只手爪抓住主轴上的刀具，同时另一手爪抓住位于机床正面用于暂时存放刀具的换刀位置上的刀具；之后机械手前伸并拔出主轴和换刀位置上的刀具；顺时针旋转 180°后，机械手回退，把新刀具插入主轴，把旧刀具放回换刀位置，然后，机械手返回到原始位置。换刀过程 5 秒即可完成。刀库能够储存许多不同刀具。由于每把刀具都有编码，因此可任意或按顺序选刀。

市场上也有不带数控系统的加工中心，但装上数控系统后可充分发挥机床的潜力。

坐标测量机

为了提高生产效率和加工质量，人们研发出各种坐标测量机。过去，人们只有在完成零件加工以后才能发现其缺陷，这样的检验技术再也不为人们接受。现在，这种从检测到预防的策略转变，从加工开始就提供实时反馈以便进行过程控制及修正。坐标测量机就是根据控制单元中的零件程序信息来检查零件加工过程中的精度。

激光切割、火焰切割和等离子切割

激光切割、火焰切割和等离子切割分别是利用高能量光束、聚集火焰、等离子弧来完成材料切削。对于不同的加工要求和材料厚度,每种加工方法都有特定的优势。

冲床和步冲轮廓机

冲床和步冲轮廓机利用冲模将金属板材冲压成型。沿一定路径连续冲压,形成一点一点咬削的效果,从而完成常规方法难以加工的复杂零件样式。成型加工也是数控冲床的典型应用。

【Reading Material】

Safety and Maintenance for CNC Machine

Safety Notes for CNC Machine Operations

Safety is always a major concern in a metal-cutting operation. CNC equipment is automated and very fast, and consequently it is a source of hazards. The hazards have to be located and the personnel must be aware of them in order to prevent injuries and damage to the equipment. Main potential hazards include: rotating parts, such as the spindle, the tool in the spindle, chuck, part in the chuck, and the turret with the tools and rotating clamping devices; movable parts, such as the machining center table, lathe slides, tailstock center, and tool carousel; errors in the program such as improper use of the G00 code in conjunction with the wrong coordinate value, which can generate an unexpected rapid motion; an error in setting or changing the offset value, which can result in a collision of the tool with the part or the machine; and a hazardous action of the machine caused by unqualified changes in a proven program.

Daily Maintenance

1. Checking the External View

(1) Machine oil (cutting oil, lubrication oil) has been scattered onto the servomotor detector or main unit of the NC, or is leaking.

(2) Damage is found on the cables of the movable blocks, or the cables are twisted.

(3) Filter clogging.

(4) A door of the control panel is not open.

(5) Ambient vibration.

2. Checking the inside of the control unit

Check that the following troubles have been eliminated:

(1) Cable connectors are loosened.

(2) Installing screws are loosened.

(3) Attachment amplifier screws are loosened.

(4) The cooling fan operates abnormally.

(5) Cable damage.

Fault Diagnosis and Action

When a running fault occurs, examine the correct cause to take proper action. To do this, execute the checks below:

1. Checking the Fault Occurrence Status

Check the following:

(1) When did the fault occur?

Time of day when the fault occurred.

(2) During what operation did the fault occur? What running mode?

Operating procedure?

Preceding and succeeding operations?

Set/display unit's screen?

During I/O operation?

What fault occurred?

What does the alarm display of the set/display unit's alarm diagnosis screen indicate?

What does the machine sequence alarm indicate?

Is the CRT screen normal?

2. Fault Example

(1) The power cannot be turned on. Check the following points:

The power is being supplied?

(2) The NC unit does not operate when being activated.

Check the following points:

Mode selected normally?

All conditions for start satisfied?

Override or manual speed = 0?

No reset signal is being generated.

No feed hold signal is being generated.

Machine lock is on.

3. Alarm Message

When the menu key ALARM is pressed, the ALARM/DIAGN screen is displayed.

(1) Alarm

The code and number or message relating to an operation alarm, program error, servo alarm, or system error are displayed.

(2) Stop code

The automatic operation disable state or stop state in automatic operation mode is displayed in code and error number.

(3) Alarm message

The alarm messages specified by the user PLC (built-in) are displayed.

(4) Operator message

The operator messages specified by the user PLC are displayed.

Unit 5   Elements of CNC Machine Tools

Text

The CNC machine tool system consists of six basic components. They are the part program, the machine control unit (MCU), the measuring system, the servo control system, the actual CNC machine tool (lathe, drill press, milling machine, etc.), and the adaptive control system. It is important to understand each element prior to actual programming of a numerically controlled part.

Part Program

The part program is a detailed set of commands to be followed by the machine tool. Each command specifies a position in the Cartesian coordinate system (x, y, z) or motion (workpiece travel or cutting tool travel), machining parameters and on/off function. Part programmers should be well versed with machine tools, machining processes, effects of process variables, and limitations of CNC controls. [1] The part program is written manually or by using computer-assisted language such as APT (Automated Programming Tool).

Machine Control Unit (MCU)

The machine control unit (MCU) is a microcomputer that stores the program and executes the commands into actions by the machine tool. The MCU consists of two main units: the Data Processing Unit (DPU) and the Control Loops Unit (CLU). The DPU software includes control system software, calculation algorithms, translation software that converts the part program into a usable format for the MCU, interpolation algorithm to achieve smooth motion of the cutter, editing of part program. [2] The DPU processes the data from the part program and provides it to the CLU which operates the drives attached to the machine lead-screws and receives feedback signals on the actual position and velocity of each one of the axes. [3] The CLU consists of the circuits for position and velocity control loops, deceleration and backlash take-up, function controls (such as spindle on/off).

In CNC systems, the DPU functions are always performed by the control program contained in the CNC computer. The major part of the CLU, however, is always implemented in the most sophisticated CNC systems.

The Measuring System

The term measuring system in CNC refers to the method a machine tool uses to move a part from a reference point to a target point. A target point may be a certain location for drilling a hole, milling a slot, or other machining operation. The two measuring systems used on CNC machine are the absolute and incremental. The absolute (also called coordinate) measuring system uses a fixed reference point (origin). It is on this point that all positional information is based. [4] In other words, all the locations to which a part will be moved must be given dimensions relating to that original fixed reference point. [5] The incremental measuring system (also called delta) has a floating coordinating system. With the incremental system, the machine establishes a new origin or reference point each time the part is moved. Notice that with this system, each new location bases its values in X and Y

from the preceding location. One disadvantage to this system is that any errors made will be repeated throughout the entire program, if not detected and corrected.

Servo Control System

CNC servomechanisms are devices used for producing accurate movement of a table or slide along an axis. Instead of causing motion by manually turning cranks and hand-wheels as is required on conventional machine tools, CNC machines allow motions to be actuated by servomotors under control of the CNC, and guided by the part program. Generally speaking, the motion type (rapid, linear, and circular), the axes to move, the amount of motion and motion rate (feed rate) are programmable with almost all CNC machine tools.

Both electric and hydraulic powers are used to achieve slide motion. There are a number of very effective, responsive, and thoroughly proved hydraulic systems currently in use. But by far the most common power source is the electric motor. Three types of motor are stepping motors, non-stepping motors and magnetic linear drive motors. Stepping motors are special type of motor designed so that they rotate in sequential finite steps when energized by electrical pulses. AC servo motors are larger than DC motors providing equivalent power, and are also more costly. However, they need less maintenance and this is a factor very much in their favor. The newest slide drive system to appear on machinery is the magnetic linear drive system. Machine movement is obtained through the activation of magnetic forces alone, using high force linear motors.

CNC machines that use an open-loop system contain no feedback signal to ensure that a machine axis has traveled the required distance. That is, if the input received is to move a particular table axis 1.000 in., the servo unit generally moves the table 1.000 in. There is no means for comparing the actual table movement with the input signal, however. The only assurance that the table has actually moved 1.000 in. is the reliability of the servo system used. Open-loop systems are, of course, less expensive than closed-loop systems.

A closed-loop system compares the actual output (the table movement of 1.000 in.) with the input signal and compensates for any errors. A feedback unit actually compares the amount the table has been moved with the input signal. [6] Some feedback units used on closed-loop systems are transducers, electrical or magnetic scales, and synchros. Closed-loop systems greatly increase the reliability of CNC machines.

A CNC command executed within the control (commonly through a program) tells the drive motor to rotate a precise number of times. The rotation of the drive motor in turn rotates the ball screw. And the ball screw drives the linear axis. A feedback device at the opposite end of the ball screw allows the control to confirm that the commanded number of rotations has taken place.

The Actual CNC Machine Tool

The machine tool could be one of the following: lathe, milling machine, coordinate measuring machine, etc. Unlike the conventional machines, the CNC machine tools have higher strength and rigidity, better stability and vibration resistance, and lower clearances.

Adaptive Control System

For a machining operation, the term adaptive control denotes a control system that measures certain output process variables and uses these to control speed and /or feed. [7] Some of the process variables that have been used in adaptive control machining system include spindle deflection or force, torque, cutting temperature, vibration amplitude, and horsepower. In other words, nearly all the metal-cutting variables that can be measured have been tried in experimental adaptive control systems. The motivation for developing an adaptive machining system lies in trying to operate the process more efficiently. The typical measures of performance in machining have been metal removal rate and cost per volume of metal removed.

The main benefits of adaptive control in machining are as followings:

- **Increased production rates.** On-line adjustments to allow for variations in work geometry, material, and tool wear provide the machine tool with the capability to achieve the higher metal removal rates that are consistent with existing conditions.
- **Increased tool life.** Adaptive control will generally provide a more efficient and uniform use of the cutter throughout its tool life. Because adjustments are made in the feed rate to prevent severe loading of the tool, fewer cutters will be broken.
- **Greater part protection.** The force limit can be established on the basis of work size tolerance. The part is protected against an out-to-tolerance condition and possible damage.
- **Less operator intervention.** The advent of adaptive control machining has transferred the control of the process even further out of the hands of the operator and into the hands of management via the part programmer.
- **Easier part programming.** The selection of feed is pretty much left to the controller unit rather than to the part programmer. The constraint limit on force, horsepower, or other variable must be determined according to the particular job and cutter used. In general, the part programmer's task requires a much less conservative approach than for numerical control. Less time is needed to program and fewer tryouts are necessary.

Technical Words

component [kəmˈpəunənt]	n.	零件；组件；部件
lathe [leið]	n.	车床
command [kəˈmænd]	n.	命令
programmer [ˈprəuˌgræmə]	n.	程序员
execute [ˈeksikjuːt]	v.	执行（指令）
algorithm [ˈælgəriðəm]	n.	算法
interpolation [inˌtəpəuˈleiʃən]	n.	插补
drive [draiv]	n.	驱动；驱动器
	v.	驱动
lead-screw	n.	导杆，丝杠
circuit [ˈsəːkit]	n.	电路
memory [ˈmeməri]	n.	存储器
slide [slaid]	n.	滑块；滑板

		v.	使滑动
servomotor	[ˈsəːvəuˌməutə]	n.	伺服电动机
motor	[ˈməutə]	n.	马达；原动力
pulse	[pʌls]	n.	脉冲
magnetic	[mægˈnetik]	adj.	磁铁的；磁性的
transducer	[trænzˈdjusə]	n.	传感器；换能器
torque	[tɔːk]	n.	转矩；力矩；扭矩

Technical Phrases

machine tool	机床
part program	工件程序
machine control unit (MCU)	机床控制装置
measuring system	测量系统
servo control system	伺服控制系统
drill press	钻床
milling machine	铣床
adaptive control system	自适应控制系统
cutting tool	切削工具
machining parameter	加工参数
process variable	工艺变量
Automated Programming Tool (APT)	自动编程工具软件
Data Processing Unit (DPU)	数据处理装置
Control Loop Unit (CLU)	回路控制装置
interpolation algorithm	插补算法
feedback signal	反馈信号
driving device	驱动设备
feedback device	反馈器件
floating coordinating system	浮点坐标系统
electric power	电力
hydraulic power	液压动力
stepping motor	步进电动机
drive motor	驱动电动机
electrical pulse	电脉冲
open-loop system	开环系统
input signal	输入信号
closed-loop system	闭环系统
ball screw	滚珠丝杠

Notes

(1) Part programmers should be well versed with machine tools, machining processes, effects of process variables, and limitations of CNC controls.

句中，Part programmers 是指编制工件加工程序的程序员；well versed with... 意为"非常熟悉……"；effects of process variables 可译为"工艺变量产生的影响"。

(2) The DPU software includes control system software, calculation algorithms, translation software *that converts the part program into a usable format for the MCU*, interpolation algorithm to achieve smooth motion of the cutter, editing of part program.

句中，that converts the part program into a usable format for the MCU 是定语从句，修饰 translation software；translation software 为"翻译软件"之意。

(3) The DPU processes the data from the part program and provides it to the CLU *which operates the drives attached to the machine lead-screws and receives feedback signals on the actual position and velocity of each one of the axes.*

本句主干成分为 The DPU **processes** the data... and **provides** it to the CLU..., 其中 DPU 是主语，两个谓语动词是 processes 和 provides。此外，which **operates** the drives... and **receives** feedback signals... 为定语从句，修饰 CLU；从句中谓语动词也有两个，即 operates 和 receives。

(4) It is on this point that all positional information is based.

本句中，"It is + n. + that 从句"属于强调语句，强调的主体是 this point，也就是上一句提及的 reference point（基准点）。

(5) In other words, all the locations *to which a part will be moved* must be given dimensions relating to that original fixed reference point.

句中，to which a part will be moved 为定语从句，修饰 all the locations。

(6) A feedback unit actually compares the amount *the table has been moved with the input signal.*

句中的 the table has been moved with the input signal 为定语从句，修饰 the amount。

(7) For a machining operation, the term adaptive control denotes a control system *that measures certain output process variables and uses these to control speed and /or feed.*

句中用于修饰 a control system 的定语从句是 that measures certain output process variables and uses these to control speed and /or feed，其中两个谓语动词分别是 measures 和 uses。

Exercises

(1) Place a "T" after sentences that are true and an "F" after those that are false.

1) The motion of a CNC machine includes the cutting tool travel and the workpiece travel.
2) The major part of the DPU is always implemented in the most sophisticated CNC systems.
3) The absolute measuring system has a floating coordinating system.
4) DC servo motors need less maintenance and this is a factor very much in their favor.
5) A closed-loop system compares the actual output with the input signal and compares the amount the table has been moved with the input signal.

(2) Fill in the blanks according to the text with the words given below. Make changes if necessary.

reference   execute   servomechanism   stability   command   adaptive

It is important to understand each element of CNC machines. The part program is a detailed set of _____ to be followed by the machine tool. The MCU is a microcomputer that stores the program and _____ the commands into actions by the machine tool. The measuring system refers to the method a machine tool uses to move a part from a _____ point to a target point. CNC _____ are devices used for producing accurate movement of a table or slide along an axis. The CNC machine tools have higher strength and rigidity, better _____ and vibration resistance, and lower clearances. The _____ control system will increase productivity by preventing or sensing damage to the cutting tool. Knowing all the functions of CNC machine tool elements will help when studying CNC machine tools.

(3) Answer the following questions:
1) How many basic components does the CNC machine tool system consist of? What are they?
2) Is the part program written manually only? Why?
3) What is the function of the lathe bed?
4) What does the DPU software include? What does the CLU consist of?
5) What is the main disadvantage of the incremental measuring system?
6) What is the difference between open-loop systems and closed-loop systems?

【参考译文】

第5课　CNC机床的组成

CNC机床系统由六部分组成，即工件加工程序、机床控制装置、坐标测量装置、伺服控制系统、实际CNC机床（车床、钻床、铣床等）以及自适应控系统。在编写零件的数控加工程序前，弄清CNC机床的每个组成部分是重要的。

工件加工程序

工件加工程序是机床必须遵循的一组详细指令。每条指令指明了在坐标系中的位置（x，y，z）、运动（工件或刀具运动）、加工参数以及开启与停止。工件加工程序员应该非常熟悉机床、加工工艺、工艺变量的影响以及CNC机床控制的局限性。工件加工程序可手工编写，也可用像APT（自动编程工具）这样的计算机辅助程序语言来编写。

机床控制装置（MCU）

机床控制装置（MCU）是一台微型计算机，其中存有加工程序，机床通过执行程序指令来完成各种操作。MCU主要由数据处理装置（DPU）、回路控制装置（CLU）组成。DPU软件包括控制系统软件、算法软件、翻译软件、插补算法及程序编辑等，其中翻译软件可将加工程序转换为MCU可用格式，插补算法可使刀具平稳运动。DPU用来处理工件程序中的数据并将结果输送到CLU，而CLU则启动装在机床丝杠上的驱动器并接受关于加工轴实际位置与进给速度的反馈信号。CLU 由以下电路组成：位置与速度控制回路、减速和后冲拉紧电路、功能控制电路（如主轴的启停控制等）。

在 CNC 系统中，DPU 的功能往往是在 CNC 计算机程序控制下完成的，而 CLU 的主要任务往往是在大多数复杂的 CNC 系统中完成。

测量系统

在 CNC 系统中，"测量系统"是机床把工件从参考点运送到目标点时所采用的程序，目标点则是指钻孔、铣槽或其他加工过程中的某一位置。CNC 机床采用绝对测量、增量测量这两种测量系统。绝对测量系统把一个固定点作为原点，所有位置数据都以此为基准；换言之，不论工件移动到哪里，其位置数据必须是工件到原点的距离。增量测量系统采用浮点坐标系统，因此每当工件移动时，机床会建立一个新原点或参考点。请注意，在此系统中，每个新位置的 X、Y 坐标都是从前一个位置计算起。这种系统的缺陷是，出现任何错误，若没有发现并纠正，会在整个加工中反复出现。

伺服控制系统

CNC 伺服机构是使工作台或滑块沿坐标轴精确运动的装置。不像传统机床那样使用手动方式操纵旋转曲柄或手轮使机床运动，CNC 机床是在 CNC 控制下用伺服电动机启动运动并由工件程序引导加工过程。一般来说，几乎所有 CNC 机床中，运动方式（快速、直线、圆弧）、加工轴移动、移动多少、进给率等都由程序控制。

滑块部件既可以用电动方式驱动，也可以用液压方式驱动。目前使用的液压装置很多，这些装置已完全证明非常有效、响应快速，不过到目前为止，最常用的动力源还是电动机。常用电动机有三种，即步进电动机、非步进电动机、永磁直线电动机。专门设计的步进电动机，在电脉冲激励下，以有限步长有序旋转。交流伺服电动机比提供等效动力的直流电动机大一些，价格也高，但交流伺服电动机便于保养维护，这是受欢迎的一个因素。机床使用的最新滑台驱动系统是"永磁线性驱动系统"，仅靠大功率直线电动机产生的磁力足以驱动机床运动。

采用开环控制的 CNC 机床，没有反馈信号来保证加工轴是否移动了所要求的距离。也就是说，如果接收的输入信号要求某个特定工作台移动 1.000 英寸，伺服装置一般会驱动工作台移动 1.000 英寸，但没有办法将工作台实际移动量与输入信号进行比较，唯一能够保障工作台移动 1.000 英寸的因素就只能是所用伺服系统的可靠性了。当然，开环系统比闭环系统价格便宜一些。

闭环系统将实际输出（工作台移动 1.000 英寸）与输入信号进行比较，并对任何误差进行补偿。反馈装置把工作台实际移动了多少与输入信号进行对比。有些闭环系统所用的反馈装置有传感器、电子或磁标尺以及同步机。采用闭环系统使 CNC 机床的可靠性大大提高。

每一条 CNC 指令都是在控制器中由程序控制执行的；CNC 指令命令驱动电动机精确转过所要求的次数；驱动电动机的转动又使得滚珠丝杠转动，滚珠丝杠又驱动直线轴运动。滚珠丝杠另一端的反馈部件使控制器确认完成指令要求的转数。

实际的 CNC 机床

实际使用的机床有以下几种：车床、铣床、坐标测量机等。与传统机床不同，CNC 机床具有较高的强度和刚度、较好的稳定性和抗振性以及较小的间隙。

自适应系统

对某一项加工操作来说，术语"自适应控制"是指通过检测某些过程变量并利用检测结果调节转速或进给量的控制系统。自适应加工所检测的过程变量有：主轴偏移量、主轴受力、主轴转矩、切削温度、振动幅度及驱动功率。换句话说，几乎所有可检测的切削变量，在自适应控制系统中都被检测过。研究自适应加工系统的动机在于设法有效控制加工过程。衡量加工性能的主要标准有两项：切削率和单位体积的切削成本。

自适应控制在机械加工中的主要优点如下：

- **生产效率提高**。工件尺寸、加工材料、刀具磨损的在线调整，使机床有能力在已有条件下提高材料切除率。
- **刀具寿命延长**。一般来说，自适应控制使刀具在整个寿命期内得以均匀而有效的运用。由于进给率可自动调节，从而防止刀具过载，刀具损坏较少。
- **工件保护程度高**。可根据工件尺寸公差确定其受力极限，保护工件免于超差及其他可能出现的损坏。
- **操作者干预少**。自适应加工的诞生，使过程控制不再过分依赖操作工，而是通过程序员的管理进行控制。
- **编程更加便捷**。进给量的选择主要由控制器完成而不是程序员。主轴受力、驱动功率及其他变量的极限值，都必须根据具体工件和所用刀具来确定。总体上讲，与数控机床编程相比，自适应控制的编程方法更为先进、编程时间更短、程序调试次数更少。

【Reading Material】

Safety Precautions of CNC Machines

In order to reduce the possibility of safety hazard and keep normal operation function, the following safety precautions are very important.

1. For Operation

To minimize the risk of accidents, follow these safety precautions:

(1) Wear safety shoes.

(2) Wear safety glasses.

(3) Wear safety hats and work clothes. Clothing should not be loose; in particular shirt cuffs should be buttoned or tight against the arm.

(4) Do not operate the machine wearing gloves.

(5) Do not touch the workpiece or spindle when machine is in operation.

(6) Sufficient light should be provided for the environment of machinery, and always keep the environment clean as far as possible.

(7) Do not use high-pressure air for blowing dust or cutting chips near the controller.

(8) A rigid floor should be provided for the operation area.

(9) The ground for the installation of machine should be rigid enough.

(10) Don't operate this machine if you feel uncomfortable.

(11) Don't leave the tool on the working table or board.

(12) Please check the condition of machine before operation. Ensure the best performance of

the machine.

(13) Every day, after turning on the machine, the machine needs to warm-up to ensure the life of spindle. The minimum time required is 30 minutes with the spindle speed set at 3000 rpm.

(14) Please stop all functions of the machine when installing the workpiece.

(15) Don't open the protective door during the operation.

2. For Machine

Before operating the machine the user needs to understand this manual very well.

(1) The operator or maintenance personnel should pay attention to all warning labels. Do not break or remove any marking label on machine.

(2) Expect when required for adjustment or repair; every door should be closed so as to keep unwanted objects away from the machine.

(3) During adjustments, repairs, or maintenance, be sure to use appropriate tools.

(4) Do not move or change the location of limit switched for the purpose of modifying the travel of machine.

(5) Please stop operating immediately and press emergency stop button, in case of any problem of machine.

(6) The following precautions should be paid attention to for daily work:

- During operation, do not place any part of your body on the movable parts, such as spindle, ATC, working table, and chip conveyor.
- Do not clean tool and/or cutting chip of working table with bare hand. Any cleaning work could be carried out only after the machine stops completely.
- Before adjustment of cutting fluid, air and fluid nozzle, the machine should be stopped completely.

(7) In normal situation, the daily work should be terminated as the following procedures:

- Press the emergency stop button.
- Turn off the power.
- Clean the working table.
- Spray anti-rust oil on the spindle and working table.

3. For Electricity

During inspection and maintenance, the following precautions should be taken:

(1) Do not hit controller by force in any case.

(2) Only use the cable, which is specified by the manufacturer. Only use the cable with appropriate length. If some of cable is laid down on the ground, it should be provided with appropriate protection.

(3) Only the manufacturer and the authorized agent are allowed to modify the setting of parameters of controller.

(4) Do not change the setting values of controller and control buttons.

(5) Do not overload the socket and conductors.

(6) Before inspect and maintenance electrical components, be sure to disconnect the device of

controller and main power, and lock it at "OFF" position.

(7) Do not use any wet tool to touch the electrical components.

(8) Only use the fuse which approves by manufacturer; the fuse with high capacity or copper wire is prohibited.

(9) Only trained and qualified personnel are allowed to open the door of electrical cabinet for maintenance, and the others are prohibited.

Unit 6　CNC Machining Process

As we all know, determining the sequence of individual manufacturing operations needed to produce a given part or product is very important before CNC machining. It is directly related to the success of the program. The better prepared the programmer, the easier and more successful the programming task will be.[1] So a good programmer must be a good process planner first.

Process sheets, also called routing sheets, are used by most manufacturing companies to specify the sequence of machining operations to be performed on a workpiece during the manufacturing process. Table 6-1 is an example sequence of operations form.

Table 6-1　Example sequence of operations form

Part number: D23957		Date: 8/16/99		Sequence of operations planning form			
Part name: Top plate		Department: 47					
Machine: MV40		Programmer: CL					
Seq.	operation	tool	station	ipm	rpm	Note	Block
1	Rough face mill top	4"face mill	1	9.17	382		
2	Rough mill outside contour	1'end mill	2	4.75	400		
3	Center drill all holes	#3 center drill	3	3.0	1800		
4	Drill (4) holes for 1/4-20	13/64 drill	4	35	1600		
5	Drill (1) 31/32 hole	31/32 drill	5	6.0	380		
6	Ream (1) 1000 hole	1000 reamer	6	8.0	350		
7	Tap (4) 1/4-20 holes	1/4-20 tap	7	22.5	450		
8	Finish face mill top	4"face mill	8	7.0	550		
9	Finish mill outside contour	1'end mill	9	4.0	500		

While developing a machining process, it is important to consider the cutting conditions you will use with each tool in the program. These cutting conditions involve functions like spindle, feed rate, depth of cut, coolant, tool material and workpiece material.

The example of material removal below demonstrates in details the procedure for machining a curved surface component, in this case a wing section.

Roughing
This is the first stage of machining where the object is to quickly remove the bulk of the waste material, normally with the aid of a ripper cutter.[2]

Semi Roughing
This stage of machining generally uses a smaller cutter than roughing, typically an end mill, although the aim is still to remove the bulk of the waste material.

Semi Finishing
This stage, using a relatively large ball nosed cutter, is to start to form the final profile of the

workpiece.

Finishing

The final stage, and the longest process of all, is the final cut to the desired size. A small ball nosed cutter traversing across the surface produces the finished shape.

Although this is the final machining stage, there is still much work to do in the form of hand polishing and finishing before the article is completed.

Besides, you have to do with the tooling used by program. Tooling problems can cause even a perfectly coded program to fail. [3] Pay attention to the types of cutter used. It is important to know the functions of various cutters during developing machining process.

Ripper Cutter

Used for the rapid removal of large amounts of metal, the serrations along the cutting edge literally rip the material away. This cutter is designed to cut along its sides only.

End Mill

Similar to a ripper only without the serrations, it is used for removing the bulk of the material. The cutting edges, like the ripper, are down the sides of the tool.

Slot Drill

Typically this cutter has less cutting surfaces than the two above, and unlike the end mill and ripper, can be used to plunge directly into the workpiece. As a result of the reduction in cutting surfaces there is better swarf removal meaning the tool can be used in enclosed pockets and slots.

Ball Nosed Slot Drill

This cutter, as its name suggests, has a semi-circular cutting face, with the aid of computer software it is possible to cut tangentially to the workpiece enabling curved surfaces to be machined. [4]

Meanwhile, the programmer is usually responsible for developing the work holding setup required to hold the workpiece during machining. Table 6-2 shows an example of a universal setup sheet. While each CNC user will develop setup sheets in a way that best suits their needs, this example shows many of the techniques commonly used.

Table 6-2 A universal setup sheet

Part number:		Operation:		Setup sheet
Part name:		Programmer:		
Machine:		Date:		
Tools				Work holding setup instructions
Station	Description		Min. len.	
				Sketch of work holding setup

Notice how the setup sheet allows cutting tools and their stations to be well described. Also notice that there is room to make a setup sketch and give the operator written instructions. The less a programmer leaves to the setup person's imagination, the less the chance there will be of a mistake being made during the setup. [5]

With the setup documentation prepared, you can easily reference fixture and clamp locations (avoiding obstructions in your program), program zero location, and the correct tool station and offset numbers.

Because of the problems encountered with manual process planning, attempts have been made in recent years to capture the logic, judgment, and experience required for this important function and incorporating them into computer programs. Based on the characteristics of a given part, the program automatically generates the manufacturing operation sequence. A computer-aided process planning (CAPP) system offers the potential for reducing the routine clerical work of manufacturing engineers. At the same time, it provides the opportunity to generate production routings which are rational, consistent, and perhaps even optimal.

Technical Words

coolant ['kulənt]	n. 冷却剂
roughing ['rʌfiŋ]	n. 粗加工
finishing ['finiʃiŋ]	n. 精加工
plunge [plʌndʒ]	v. 插入
slot [slɔt]	n. 槽
fixture ['fikstʃə]	n. 夹具；固定装置
automatically [ˌɔtə'mætikli]	adv. 自动地；机械地
machine [mə'ʃin]	n. 机器
	v. 机械加工

Technical Phrases

Computer Numerical Control (CNC)	计算机数字控制
cutting condition	切削状态
feed rate	进给率
ripper cutter	纵切刀
end mill	立铣刀，端铣刀
ball nosed cutter	球头铣刀
slot drill	槽钻
cutting surface	切削表面
ball nosed slot drill	球头槽钻
work holding setup	工件夹持装置
cutting tool	刀具
computer-aided process planning (CAPP)	计算机辅助工艺规划

Notes

(1) The better prepared the programmer, the easier and more successful the programming task will be.

这是一个比较句,意思是:"程序员把加工工序准备得越好,编制程序就越轻松、顺畅。"

(2) This is the first stage of machining *where the object is to quickly remove the bulk of the waste material*, normally with the aid of a ripper cutter.

句中,where the object is to quickly remove the bulk of the waste material 是定语从句,修饰 the first stage of machining; 其中的 waste material 可引申为"冗余材料"。

(3) Tooling problems can cause even a perfectly coded program to fail.

句中,cause 意为"引起",perfectly coded program 意为"完美的编码程序"。

(4) This cutter, as its name suggests, has a semi-circular cutting face, with the aid of computer software it is possible to cut tangentially to the workpiece enabling curved surfaces to be machined.

文中,as its name suggests 可译为"顾名思义"; semi-circular 意思是"半圆形"; with the aid of 意思是"借助于"。

(5) The less a programmer leaves to the setup person's imagination, the less the chance there will be of a mistake being made during the setup.

这是一个比较句,意思是:"程序员为制订作业准备的人留下的想象空间越小,作业准备过程中出错的可能性就越小"。

Exercises

(1) Place a "T" after sentences that are true and an "F" after those that are false.

1) The sequence of individual manufacturing operations needed to produce a given part or product is directly related to the success of the program.

2) The machine maker is usually responsible for developing the work holding setup.

3) The end mill is used for removing the bulk of the material.

4) Ripper cutter is used for the rapid removal of large amounts of metal designed to cut along its sides only.

5) Roughing is to start to form the final profile of the workpiece.

(2) Fill in the blanks according to the text with the words given below. Make changes if necessary.

shape   process   setup   profile   cutter

The object of roughing and semi roughing is to quickly remove the bulk of waste material, but semi roughing generally uses a smaller _____ than roughing. Semi finishing forms the final _____ of the workpiece, while finishing produces the finished _____. Besides, pay attention to the types of cutter used. It is important to know the functions of various cutters during developing machining _____. Meanwhile, the programmer is usually responsible for developing the work holding _____ required to hold the workpiece.

(3) Answer the following questions:
1) Why do we usually say a good programmer must be a good process planner first?
2) What is the function of the process sheets?
3) What do the cutting conditions involve?
4) What is the main procedure for machining a curved surface component?
5) What is the function of a setup sheet?
6) What is CAPP? Tell its main functions.

【参考译文】

第6课　CNC 加工工艺

众所周知，在 CNC 加工前确定加工工件或产品所需要的工序是非常重要的，因为它直接关系到程序能否顺利运行。程序员把加工工序准备得越好，编制程序就越轻松、顺畅。因此，好的程序员首先必须是一位好工艺师。

绝大多数制造公司使用工艺单（也叫流程单）来说明工件加工过程所用加工操作的顺序。表6-1 所示是一个典型的工艺单。

编制加工工艺时，应考虑程序控制加工中每个刀具的切削条件，这一点很重要；这些切削条件涉及主轴、进给率、切削深度、冷却、刀具材料以及工件材料等。

下面举例详细说明加工曲面工件（机翼部分）时切削材料的过程。

粗加工

这是加工的第一阶段，其目的是快速切掉大量冗余材料，通常用纵切刀完成。

半粗加工

这一阶段使用的刀具比粗加工要小，一般使用端铣刀，但加工目的仍然是切除大量冗料。

半精加工

这一步的目标是用比较大的球头铣刀开始形成工件的最终轮廓。

精加工

精加工是最后一个阶段，也是所有加工中最长的阶段，它的任务是把工件加工到期望的尺寸精度。精加工采用小型球头铣刀在工件表面移动，从而加工出最终的形状。

虽然这是最后的加工阶段，但在加工出完整的工件之前仍然有很多手工抛光的精加工工作要做。

除此以外，还应重视程序中使用的刀具。刀具存在问题会使完美的编码程序在加工时失败。用什么类型的刀具，应予以关注。在制订加工工艺期间，弄清楚各种刀具的功能是很重要的。

纵切刀

纵切刀用锯齿状的切削刃来快速切除大量金属。这种刀具在设计上只能沿其边缘进行切削。

端铣刀

端铣刀与纵切刀类似,但没有锯齿,可用来大量切除金属。与纵切刀一样,端铣刀的切削刃也在刀具两侧。

槽钻

槽钻的切削面一般比以上两种刀具少,而且与端铣刀和纵切刀不同的是,槽钻可直接插进工件里面。由于切削面少,因而排屑较好,这意味着槽钻可以在密闭容器或狭缝中使用。

球头槽钻

顾名思义,球头槽钻的切削面呈半圆形,在计算机软件辅助下可沿切线方向切削工件以加工出各种曲面。

与此同时,加工期间夹持工件所需要的作业单一般由程序员负责制订。表6-2 是一份通用的工件夹持作业准备单。虽然每位 CNC 操作者会制订出最适合他们需要的作业准备单,但表中展现了许多常用的技巧。

请注意作业准备单是如何充分描述刀具及刀位的。此外还应注意留有空间,以便绘制作业准备草图并给操作者留下书面说明。程序员为制订作业准备的人留下的想象空间越小,在作业准备过程中出错的可能性就越小。

编好作业准备文件,就可方便地确定夹具位置(以免给程序带来干扰)、程序零点位置、合适的刀位以及刀具偏置号。

过去,人工编制工艺规划时常遇到一些问题;近年来,人们针对工艺规划中的重要功能进行尝试,试图探寻其中的逻辑、判断及与这些重要功能相关的经验,并将其融入到计算机程序当中;现在,计算机程序可以根据工件特点自动生成加工工序。计算机辅助工艺规划系统的出现,有望减轻制造工程师们的工作量。与此同时,使用计算机辅助工艺规划形成的生产工艺,不但合理性与一致性好,而且有可能是最佳工艺。

【Reading Material】

Controlling in the CNC Machining Process

After the design has been translated into machine language through G-Code, it is now time to put the G-Code and the instructions into motion.

The CNC control stage of the CNC machining process is broken down into three parts namely the CNC control computer, the CNC control software, and the CNC controller itself.

CNC Control Computer

The first part of the CNC control stage in the CNC machining process is the CNC control computer. The CNC control computer doesn't have to be the newest or fastest computer out there.

There are many CNC computers that are mere 386's or 486's. If you aren't familiar with these 86's then simply think of them as the computers of the mid nineties. They were the predecessors of the Pentium chips from Intel.

These CNC computers don't have to be fast, as you see. It is more likely to run one program anyway. It won't be the type of computer that has to multitask like Windows. The CNC control computer of the CNC machining process will basically be used to read the G-Code that was written in the CAM stage of the CNC machining process. One other thing that you have to put in mind is that the

CNC control computer is going to be in a rugged environment.

Expect your CNC control computer to be covered in dust because they will be near your work environment. They are usually beside the machines that spew dirt and dust when products are being manufactured.

The CNC control computers for the CNC machining process will cost you around fifty to a thousand dollars depending on the computer. If you're getting the low end, prehistoric computers then spare change in your pocket would do.

CNC Control Software

The CNC control software in the CNC machining process is the software that will read the G-Code program. You've got the hardware, now it's time to get your hands on the software that will run your G-Code.

Some of the well-known CNC control software packages out there are the Mach 3 Artsoft, Turbo CNC, and the CNC Pro. The control software packages used in CNC machining are usually not so demanding when it comes to processor power and memory usage. Control software costs can range from totally free to about a thousand dollars.

There is freeware out there and there is the highly customized CNC control software. On a ballpark estimate, the average CNC control software used in CNC machining like the Mach 3 will cost you around one hundred fifty dollars.

CNC Controller

The last part of the CNC controlling process is the CNC controller. The CNC Controller in the CNC machining process is the part that processes the commands from the control computer and the control software environment and then translates it into actual motion.

There are many types of CNC controllers out there. CNC controllers can vary in the numbers of axis options. They can vary in the number of accessories too. Some will usually come with motor kits and cables. Normally, CNC controllers are purchased but sometimes if you are computer savvy, you can make your own CNC controller for your CNC machining hobby. Otherwise, it is on the CNC controller where you'll be spending your money.

Depending on the number of axis and additional accessories that come with it, you'll be spending around five hundred to five thousand dollars.

Be prepared to burn your wallet on this kind of CNC machining equipment.

Unit 7　CNC Programming

Text

In an NC (Numerically Controlled) machine, the tool is controlled by a code system that enables it to be operated with minimal supervision and with a great deal of repeatability.[1] CNC (Computerized Numerical Control) is the same type of operating system, with the exception that the machine tool is monitored by a computer.

The same principles used in operating a manual machine are used in programming an NC or a CNC machine. The main difference is that instead of cranking handles to position a slide to a certain point, the dimension is stored in the memory of the machine control once[2]. The control will then move the machine to these positions each time the program is run.

The operation of the VF-Series Vertical Machining Center requires that a part program be designed, written, and entered into the memory of the control. The most common way of writing part programs is off-line, that is, away from the CNC in a facility that can save the program and send it to the CNC control[3]. The most common way of sending a part program to the CNC is via an RS-232 interface. The HAAS VF-Series Vertical Machining Center has an RS-232 interface that is compatible with most existing computers and CNCs[4].

In order to operate and program a CNC machine, a basic understanding of machining practices and knowledge of math are necessary. It is also important to become familiar with the control console and the placement of the keys, switches, displays, etc.

Before you can fully understand CNC, you must first understand how a manufacturing company processes a workpiece that will be produced on a CNC machine[5]. The following items form a fairly common and logical sequence of tasks done in CNC programming. The items are only in a suggested order, offered for further evaluation.

- Obtain or develop the part drawing.
- Decide what machine will produce the part.
- Decide on the machining sequence.
- Choose the tooling required.
- Do the required math calculations for the program coordinates.
- Calculate the speeds and feeds required for the tooling and part material.
- Write the NC program.
- Prepare setup sheets and tool lists.
- Send the program to the machine.
- Verify the program.
- Run the program if no changes are required.

Manual programming (without a computer) has been the most common method of preparing a part program for many years. The latest CNC controls make manual programming much easier than ever before by using fixed or repetitive machining cycles, variable type programming, graphic tool motion simulation, standard mathematical input and other timesaving features[6]. The need for improved efficiency and accuracy in CNC programming has been the major reason for development of a variety of methods that use a computer to prepare part programs.[7] Computer assisted CNC programming has been around for many years.

Technical Words

supervision [ˈsjupəˈviʒən]	n. 监督，管理
monitor [ˈmɔnitə]	n. 监视器
	v. 监视
memory [ˈmeməri]	n. 存储器
interface [ˈintəfeis]	n. 界面；接口；接口电路
compatible [kəmˈpætəbl]	adj. 兼容的
display [diˈsplei]	n. 显示；显示器
	v. 显示
feed [fiːd]	n. 馈送，供给；进给
console [kənˈsəul]	n. 控制台
sheet [ʃiːt]	n. 清单
verify [ˈverifai]	v. 校验
graphic [ˈgræfik]	adj. 图形的
simulation [ˌsimjuˈleiʃən]	n. 仿真

Technical Phrases

control console	控制面板
part drawing	零件图
machining sequence	加工顺序
program coordinate	程序坐标
setup sheet	作业准备单
tool list	刀具清单
manual programming	手工编程
off-line	离线
fixed cycle	固定循环
variable type programming	变量编程

Notes

(1) In an NC (Numerically Controlled) machine, the tool is controlled by a code system *that enables it to be operated with minimal supervision and with a great deal of repeatability.*

句中的 that enables it to be operated with minimal supervision and with a great deal of repeatability 为定语从句，修饰 a code system；a code system 可译为"编码系统"。

(2) The main difference is that instead of cranking handles to position a slide to a certain point, the dimension is stored in the memory of the machine control once.

根据上下文来看，本句中的 The main difference 指的是 NC、CNC 机床与手动机床之间的差别；once 可引申为"早已，事先"等。

(3) The most common way of writing part programs is off-line, that is, away from the CNC in a facility *that can save the program and send it to the CNC control*.

句中的 that can save the program and send it to the CNC control 为定语从句，修饰 a facility。

(4) The HAAS VF-Series Vertical Machining Center has an RS-232 interface *that is compatible with most existing computers and CNCs*.

句中的 that is compatible with most exiting computers and CNCs 为定语从句，修饰 an RS-232 interface；compatible with... 意思是"与……兼容"。

(5) Before you can fully understand CNC, you must first understand how a manufacturing company processes a workpiece *that will be produced on a CNC machine*.

句中的 how a manufacturing company processes a workpiece 是谓语动词 understand 的宾语从句，而 that will be produced on a CNC machine 是定语从句用来修饰 a workpiece。

(6) The latest CNC controls make manual programming much easier than ever before by using fixed or repetitive machining cycles, variable type programming, graphic tool motion simulation, standard mathematical input and other timesaving features.

句中的 make... much easier than ever before 意思是"使……比以往任何时候都更容易"；根据文章内容，可将 other timesaving features 引申为"其他简捷功能"。

(7) The need for improved efficiency and accuracy in CNC programming has been the major reason for development of a variety of methods *that use a computer to prepare part programs*.

这个句子较长，主语是 The need for improved efficiency and accuracy in CNC programming；that use a computer to prepare part programs 是定语从句，修饰 methods。

Exercises

(1) Place a "T" after sentences that are true and an "F" after those that are false.

1) RS-232 interface is compatible with most exiting computers and CNCs.

2) Manual programming (without a computer) has been the most common method of preparing a part program for many years.

3) In order to operate and program a CNC machine, a basic understanding of machining practices and a working knowledge of math is necessary.

(2) Fill in the blanks according to the text with the words given below. Make changes if necessary.

dimension   crank   familiar   programming   simulation

1) The main difference is that instead of _____ handles to position a slide to a certain point, the _____ is stored in the memory of the machine control once.

2) It is also important to become _____ with the control console and the placement of the keys, switches, displays, etc.

3) The latest CNC controls make manual _____ much easier than ever before by using fixed or repetitive machining cycles, variable type programming, graphic tool motion _____, standard mathematical input and other timesaving features.

【参考译文】

第7课　CNC编程

在NC机床中，刀具由编码系统控制，因而加工过程中人工监管很少、加工操作的可重复性也很好。CNC机床操作系统与NC机床相同，不同的是CNC机床是由计算机控制的。

NC机床或CNC机床编程原理与手动机床是一样的，主要区别在于手动机床通过操作手柄使滑板移到指定点，而NC或CNC机床则是将尺寸信息事先储存在机床存储器中，程序运行时控制器将控制机床移动到指定位置。

要运行VF系列立式加工中心，就需要设计和编写工件加工程序，并将其输入到控制器的存储器中。工件程序通常采用离线方式编写，也就是说脱离数控设备编程，编好后保存并传输到CNC控制器中，最常用的程序传输方式是RS-232接口。HAAS VF系列立式加工中心配有RS-232接口电路，该接口与目前大多数计算机和CNC机床都是兼容的。

为了操作CNC机床并能对其进行编程，不仅要对机床加工和数学知识有所了解，还要对控制台以及键盘、开关、显示器的布置都非常熟悉。

要想完全弄清楚CNC机床，必须首先弄清楚制造厂是如何用数控机床完成工件加工的。下面列出的条目是编制CNC程序时通常要考虑的一系列项目，这些项目的逻辑顺序仅仅是为了进一步评估而推荐的顺序。

- 获取或绘制零件图。
- 选择加工机床。
- 确定加工工序。
- 选取加工刀具。
- 计算编程坐标。
- 根据刀具和工件材料计算转速和进给量。
- 编写加工程序。
- 准备作业准备单和刀具清单。
- 把加工程序发送到机器。
- 校验加工程序。
- 如无需更改，运行加工程序。

多年来，手工编程一直是编制工件加工程序时最常用的方法。最新CNC控制器兼有固定加工循环、重复加工循环、变量编程、刀具运动图形仿真、标准数学输入以及其他简捷功能，使手工编程比以往方便了很多。为进一步提高CNC编程效率与精度，人们开发了多种使用计算机编制工件程序的方法；这种计算机辅助编程方法多年来也得到了广泛应用。

[Reading Material]

Programming with Codes

A program is written as a set of instructions given in the order they are to be performed. The instructions, if given in English, might look like this:

LINE#1 = SELECT CUTTING TOOL.

LINE#2 = TURN THE SPINDLE ON AND SELECT THE RPM.

LINE#3 = TURN THE COOLANT ON.

LINE#4 = RAPID TO THE STARTING POSITION OF THE PART.

LINE#5 = CHOOSE THE PROPER FEED RATE AND MAKE THE CUT(S).

LINE#6 = TURN OFF THE SPINDLE AND THE COOLANT.

LINE#7 = RETURN TOOL TO HOLDING POSITION AND SELECT NEXT TOOL.

But our machine control understands only these messages when given in machine code.

Before considering the meaning and the use of codes, it is helpful to lay down a few guidelines:

(1) Codes come in groups. Each group has an alphabetical address. The universal rule here is that except for G-code and macro calls, codes with the same alphabetical address cannot be used more than once on the same line.

(2) G-codes come in groups. Each G-code group has a specific group number. G-codes from the same group cannot be used more than once on the same line.

(3) There are modal G-code groups which, once established, remain effective until replaced with another code from the same group.

(4) There are non-modal G-code groups which, once called, are effective only in the calling block, and are immediately forgotten by the control.

Unit 8 Servo Controls

Text

Servo controls can be any group of electrical, hydraulic, or pneumatic devices which are used to control the position of machine tool slides. The most common servo control systems in use are the open-loop and the closed-loop systems.

In the open-loop system, the tape is fed into a tape reader which decodes the information punched on the tape and stores it briefly until the machine is ready to use it. The tape reader then converts the information into electrical pulses or signals. These signals are sent to the control unit, which energizes the servo control units. The servo control units direct the servomotors to perform certain functions according to the information supplied by the tape. The amount each servomotor will move depends upon the number of electrical pulses it receives from the servo control unit. Precision leadscrews, usually having 10 threads per inch (tpi), are used on NC machines. If the servomotor connected to the leadscrew receives 1,000 electrical pulses, the machine slide will move 1 in. (25.4mm). Therefore, one pulse will cause the machine slide to move 0.001 in. (0.0254mm). The open-loop system is fairly simple; however, since there is no means of checking whether the servomotor has performed its function correctly, it is not generally used where an accuracy greater than 0.001in. (0.0254mm) is required.

The open-loop system may be compared to a gun crew that has made all the calculations necessary to hit a distant target but does not have an observer to confirm the accuracy of the shot.

The closed-loop system can be compared to the same gun crew that now has an observer to confirm the accuracy of the shot. The observer relays the information regarding the accuracy of the shot to the gun crew, which then makes the necessary adjustments to hit the target.

The closed-loop system is similar to the open-loop system with the exception that a feedback unit is introduced into the electrical circuit. This feedback unit, often called a transducer, compares the amount the machine table has been moved by the servomotor with the signal sent by the control unit. The control unit instructs the servomotor to make whatever adjustments are necessary until both the signal from the control unit and the one from the servo unit are equal. In the closed-loop system, 10,000 electrical pulses are required to move the machine slide 1in. (25.4mm). Therefore, on this type of system, one pulse will cause a 0.0001 in. (0.00254 mm) movement of the machine slide. Closed-loop NC systems are very accurate because the command signal is recorded, and there is an automatic compensation for error. If the machine slide is forced out of position due to cutting forces, the feedback unit indicates this movement and the machine control unit (MCU) automatically makes the necessary adjustments to bring the machine slide back to position.

Technical Words

electrical [i'lektrikəl]　　　　　　　　　　　adj. 电的；用电的

hydraulic [haiˈdrɔːlik]	adj.	水的；液压的
pneumatic [njuːˈmætik]	adj.	风力的；气动的
decode [ˌdiːˈkəud]	v.	解码，译码
convert [kənˈvəːt]	v.	变换，转换
signal [signl]	n.	信号
	v.	发信号
send [send]	v.	发送
energize [ˈenədʒaiz]	v.	提供能量
direct [diˈrekt, daiˈrekt]	v.	指挥；命令
	n.	指挥；命令
receive [riˈsiːv]	v.	接收
precision [priˈsiʒən]	n.	精度
leadscrew [ˈliːdskruː]	n.	丝杠
thread [θred]	n.	（螺纹的）线数
relay [ˈriːlei]	v.	传送
	n.	继电器
transducer [trænzˈdjuːsə]	n.	传感器
instruct [inˈstrʌkt]	n.	指令
	v.	命令
command [kəˈmɑːnd]	n.	命令
	v.	命令
compensation [ˌkɔmpenˈseiʃən] (for)	n.	补偿

Technical Phrases

servo control	伺服控制
servo control unit	伺服控制单元
electrical circuit	电路
electrical device	电气设备
pneumatic device	气动设备
open-loop system	开环系统
closed-loop system	闭环系统
feedback unit	反馈单元
cutting force	切削力

Exercises

(1) Fill in the blanks according to the text with the words given below. Make changes if necessary.

compensation   electrical   open-loop   pulse   transducer

1) The three types of devices used in servo control systems are pneumatic, hydraulic, and

_____ ones.

2) According to Paragraph 2, one of the functions a tape reader has is to convert the information punched on a tape into electrical _____ or signals.

3) The _____ system is not generally used where the accuracy is greater than 0.001in.

4) The feedback unit in a closed-loop system is often called a (an) _____.

5) Closed-loop NC systems have the function of automatic _____ for error.

(2) Answer the following question:

What is the difference between the open-loop and the closed-loop systems?

(3) Explanation:
CAM　　CIM　　OEM　　CPU　　RAM　　MCU　　BCD　　EIA　　CAD
CRT

【参考译文】

第 8 课　伺服控制系统

伺服控制设备是由电力、液压与气动装置组合而成的，其作用是控制机床工作台的运动。最常用的伺服控制系统有开环与闭环两种。

在开环系统中，纸带阅读机译出穿孔带上的信息，并将其暂存起来，以备随时使用；纸带阅读机将这些信息转换成电脉冲或信号并送至控制装置，以驱动伺服控制设备；伺服控制设备按照纸带信息驱动伺服电动机，实现特定的功能。各伺服电动机的转动量取决于伺服控制单元发出电脉冲的数目。NC 机床通常采用 10 线螺纹/英寸的精密丝杠。如果与丝杠相联的伺服电动机收到 1000 个电脉冲，则工作台移动 1 英寸（25.4mm），因而一个脉冲使工作滑台移动 0.001 英寸（0.0254mm）。开环控制系统相当简单，但由于无法检测伺服电动机是否正确执行其功能，因而精度高于 0.001 英寸（0.0254mm）的场合通常并不用开环控制。

开环系统好比一位射手，虽然他做了详尽计算欲命中远程目标，但却没有观察者确认射击是否准确。

闭环系统也可比作同一位射手，但现在有一个观察者来确认射击的准确性，他将有关射击准确性信息传递给射手，射手再做必要调整，以便击中目标。

除了电路中加入反馈装置以外，闭环系统与开环系统在其他方面是相似的。反馈装置通常叫做传感器，能将伺服电动机驱动工作台移动的量与控制装置发出的信号进行比较。控制装置使伺服电动机不断作出必要调整，直到来自控制装置的信号与伺服装置的信号相等为止。在闭环系统中，10 000 个电脉冲才能使工作台移动 1 英寸（25.4mm），因而该系统的一个脉冲使工作台仅移动 0.0001 英寸（0.00254mm）。闭环数控之所以控制精度很高，就在于它能记录指令信号，并能自动补偿误差。如因切削力而使工作台位置发生偏移，反馈装置会显示这一偏移，机床控制单元（MCU）会自动进行必要调整，从而使机床工作台返回原位。

[Reading Material]

How the Control Moves the Machine

Acceleration and deceleration is what the machine does when it is changing speed. Acceleration is what the machine does when the speed is increasing and deceleration is what the machine does when it is slowing down. The machine cannot change speed instantly so that a change of speed occurs over some amount of time and a distance.

Changes of speed affect how the control moves for both rapid motion and feed motion. Rapid motion occurs independently for each axis in motion and uses acceleration set for each axis. Feed motion coordinates one or more axes to accelerate in unison, move in unison, and decelerate in unison. This type of feed motion is called acceleration before interpolation and uses a fixed acceleration rate for all axes.

Rapid motion uses constant acceleration and deceleration with maximum acceleration and maximum speed set as a parameter per axis. End point arrival in a rapid motion occurs with S-curve velocity to prevent shock vibration to the machine. A rapid motion followed by a rapid motion is blended with a rounded corner controlled by a parameter called "In Position Limit". It is usually about 0.06 inches. A rapid motion followed by a feed motion or a rapid motion in "Exact Stop" mode will always decelerate to an exact stop before the next motion.

S-curve velocity control refers to the rate of change of the acceleration or deceleration. Without S-curve, the deceleration can make abrupt changes and result in machine vibration. With S-curve at the end of a rapid move, there are only gradual changes to deceleration and thus machine vibration is reduced.

A feed motion coordinates or "interpolates" the motion of multiple axes. Feed motions always use constant acceleration before interpolation. With the Haas control, up to five axes can be in motion in a feed. These are X, Y, Z, A, and B. Maximum feed rate in a linear motion is 500 inches per minute for the linear (XYZ) axes and 300 degrees per minute for the rotary axes (AB).

Unit 9 MCU and CPU

Text

The MCU (Fig. 9-1) is the intermediary in the total NC operation. Its main function is to take the part program and convert this information into a language that the machine tool can understand so that it can perform the functions required to produce a finished part. [1] This could include turning relays or solenoids ON or OFF and controlling the machine tool movements through electrical or hydraulic servomechanisms.

DATA Decoding and Control

One of the first operations that the MCU must perform is to take the binary-coded data (BCD) from the punched tape and change it into binary digits. This information is then sent to a holding area, generally called the buffer storage, of the MCU. The purpose of the buffer area is to allow the information or data to be transferred faster to other areas of the MCU. If there were no buffer storage, the MCU would have to wait until the tape reader decoded and sent the next set of instructions. This would cause slight pauses in the transfer of information, which in turn would result in a pause in the machine tool motion and cause tool marks in the workpiece. MCUs which do not have buffer storage must have high-speed tape readers to avoid the pauses in transferring information and the machining operation.

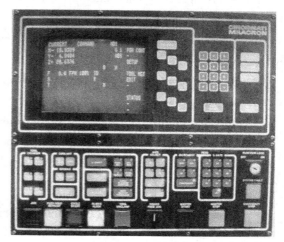

Fig. 9-1 MCU

The data decoding and control area of the MCU processes information which controls all machine tool motions as directed by the punched tape. This area also allows the operator to stop or make changes to the program manually, through the control panel.

MCU Development

Since the early 1950s, MCUs have developed from the bulky vacuum tube units to today's computer control units, which incorporate the latest microprocessor technology. Until the early 1970s, all MCU functions—such as tape format recognition, absolute and incremental positioning, interpolation, and code recognition—were determined by the electronic elements of the MCU. [2] This type of MCU was called hard-wired because the functions were built into the computer elements of the MCU and could not be changed.

The development of soft-wired controls in the mid-1970s resulted in more flexible and less costly MCUs. Simple types of computer elements, and even minicomputers, became part of the MCU. The functions that were locked in by the manufacturer with the early hard-wired systems are now included in the computer software within the MCU. This computer logic has more capabilities, is less expensive, and at the same time can be programmed for a variety of functions whenever required.

CPU

The CPU of any computer contains three sections: control, arithmetic-logic, and memory. The CPU control section is the workhorse of the computer. Some of the main functions of each part of the CPU are as follows:

1. *Control section*

1) Coordinates and manages the whole computer system.

2) Obtains and decodes data from the program held in memory.

3) Sends signals to other units of the NC system to perform certain operations.

2. *Arithmetic-logic*

1) Makes all calculations, such as adding, subtracting, multiplying, and counting as required by the program.

2) Provides answers to logic problems (comparisons, decisions, etc.)

3. *Memory*

1) Provides short-term or temporary storage of data being processed.

2) Speeds the transfer of information from the main memory of the computer.

3) Has a memory register which provides a specific location to store a word and/or recall a word.

Technical Words

solenoid [ˈsəulinɔid]	n.	螺线管
buffer [ˈbʌfə]	n.	缓冲器
pause [pɔːz]	n.	暂停
	v.	暂停
panel [ˈpænl]	n.	面板
incorporate [inˈkɔːpəreit]	v.	合并；组合
format [ˈfɔːmæt, -mɑːt]	n.	形式；格式
	v.	对……格式化
absolute [ˈæbsəluːt]	adj.	绝对的
incremental [ˌinkriˈmentəl]	adj.	增量的
interpolation [inˌtəːpəuˈleiʃən]	n.	插补
recall [riˈkɔːl]	v. & n.	调用

Technical Phrases

turn on	打开
turn off	关闭
binary-coded data	二进制编码数据
binary digit	二进制数字

buffer storage	缓冲存储器
buffer area	缓冲区
tool mark	刀痕
high-speed tape reader	高速读带机
control panel	控制面板
vacuum tube	真空管

Notes

(1) Its main function is to **take** the part program and **convert** this information into a language *that the machine tool can understand* so that it can perform the functions required to produce a finished part.

主句有两个并列表语：to take... 与（to） convert...；定语从句 *that the machine tool can understand* 修饰 a language。

(2) Until the early 1970s, **all MCU functions** — such as tape format recognition, absolute and incremental positioning, interpolation, and code recognition — **were determined by the electronic elements of the MCU.**

这是一个简单句，其结构是：all MCU functions were determined by...。

Exercises

(1) Answer the following questions：
1) What is the main function of a MCU?
2) What is the purpose of a buffer storage?
3) What consequences would result without a buffer storage in a MCU?
4) How many sections does a CPU contain? What are they?

(2) Translate the following sentences into Chinese：
1) With no buffer storage, the MCU would have to wait until the tape reader decoded and sent the next set of information.
2) Control section of a CPU can obtain and decode data from the program held in memory.

【参考译文】

第 9 课　MCU 与 CPU

MCU（见图 9-1）是整个数控机床运行的中枢环节，其主要任务是接收零件程序并转换为机床可以识别的语言，实现加工成品所需的各种功能。这就要求能用电气与液压伺服设备正确控制继电器开关状态及机床运行过程。

数据译码与控制设备

从穿孔带获得 BCD 码并转换为二进制数，这是 MCU 的首要操作之一。下一步是把这些信息送到 MCU 的保存区，通常称为缓冲存储区。缓冲区的任务是把这些信息与数据迅速传至 MCU 的其他部分。如果没有缓存区，MCU 只有等待读带机完成译码并发出下一组指令才能动作，这就会中断信息传输，使机床暂停运行、工件上留有刀痕。没有缓存区的 MCU 应配有高速读带机，以免信息传输与机床运行中出现中断现象。

MCU 的数据译码与控制设备可处理所有机床运动的信息。该设备也允许操作者用控制面板暂停机床运行及手动修改程序。

MCU 的发展

自从 20 世纪 50 年代初期以来，MCU 已从笨重的真空管装置发展成为今天的计算机控制单元，其中已经融进了最新的微处理器技术。20 世纪 70 年代初期以前，MCU 的各项功能，如纸带格式识别、绝对定位、增量定位、插补、代码识别等，均由 MCU 的电子装置实现。这种 MCU 曾称作硬控制器，因为这些功能已经固化在 MCU 的计算机部件中而无法改变。

20 世纪 70 年代中期，软控制器的发展使 MCU 更具柔性且成本更低。计算机部件，甚至小型计算机也只是 MCU 的组成部分。过去曾由制造商固化在硬件电路中的那些功能，现在都已包含在 MCU 软件中。这种计算机的逻辑能力更强、价格更低，并能对多种功能进行编程。

CPU

任何计算机的 CPU 均由三部分组成：控制器、算术-逻辑单元及存储器。控制器是主要组成部分。CPU 各组成部分的主要功能如下：

1. 控制器

1）协调、管理整个计算机系统。

2）从存储器中获取程序数据并完成译码。

3）向 NC 系统其他单元发送信号，执行一些特定操作。

2. 算术-逻辑单元

1）完成程序要求的加、减、乘、计数等运算。

2）完成逻辑处理（比较、判断等）。

3. 存储器

1）短期或临时存储正在处理的数据。

2）提高计算机主存信息的传输速度。

3）有一个存储器寄存器，提供一个特定位置来存储、调用一个字。

Unit 10　CNC

Text

In 1970, a new term—computerized numerical control (CNC)—was introduced to the NC vocabulary. The development of a CNC system was made economically possible by electronic breakthroughs which resulted in lower costs for minicomputers.

The physical components of CNC units are the same regardless of the type of machine tool being controlled. On soft-wired units, it is not the MCU but the executive program, or load tape, which is loaded into the CNC computer's memory by the control manufacturer. It is this executive program that makes the control unit "think" like a turning center or a machining center. Therefore, if there is a need to change the functions of a soft-wired or CNC unit, the executive program can be changed to suit. Usually it is the manufacturer, and not the user, who makes the change.

There have been great developments in MCUs with the introduction of large-scale ICs and microprocessors. Many features not available a few years ago give much greater flexibility and productivity to NC systems. As developments in control units continue, new features will become available which will make the machines they control more productive.

The CNC machine control unit of today has several features (Fig. 10-1) which were not found on the pre-1970 hard-wired control units.

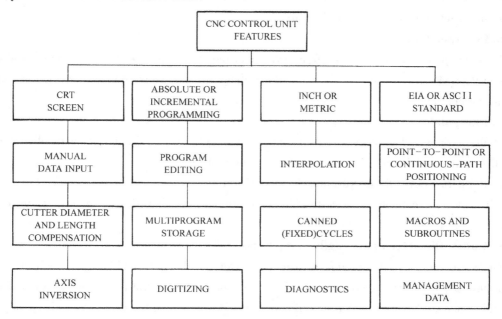

Fig. 10-1　The CNC units of today contain many features

1. Cathode-Ray Tube (CRT)

The CRT is like a TV screen which serves a number of purposes:

a. It shows the exact position of the machine table and/or the cutting tool at every position while a part is being machined.

b. The entire program for a part can be displayed on the screen for editing or revision purposes.

c. The screen assists in work setups, and some models receive messages from sensors which indicate machine or control problems.

2. Absolute and Incremental Programming

By using the proper address code G90 (absolute) or G91 (incremental), most CNC units will automatically program in that particular mode. Some of the newer CNC units are capable of handling mixed data (absolute and incremental) in a given block of data.

3. Inch or Metric

Most CNC units are capable of working in inch or metric dimensions. Either a switch on the control unit or a specific code on the part program (G70 for inch and G71 for metric) will determine the measurement system used when machining a part.

4. EIA or ASCII Code

Many of the newer CNC units will read either the Electronic Industries Association (EIA) or the American Standard Code for Information Interchange (ASCII) standard code. The control unit identifies each one by the parity check: odd for the EIA code and even for the ASCII Code.

5. Manual Data Input

Most CNC units provide some method of making changes to the part program, if necessary. This may be necessary because changes were made to the part specifications, to correct an error, or to change the machining sequences of the part.

6. Program Editing

Very few part programs are free from error from the start, and the flaws generally show up on the shop floor. Program editing is a feature which allows the part program to be corrected or changed right at the control unit.

7. Interpolation

While early models of control units were capable of only linear, circular, or parabolic interpolation, the newer models include helical and cubic interpolation.

8. Point-to-Point OR Continuous-Path Positioning

All MCUs are capable of point-to-point or continuous-path positioning or any combination of each.

9. Cutter Diameter and Length Compensation

On newer MCUs, it is possible to manually enter the diameter and/or the length of a cutter which may vary from the specification of the part program. The control unit automatically calculates what adjustments are necessary for the differences in size and moves the necessary slides to adjust for the difference. [1]

10. Program Storage

The newer CNC units generally provide large-capacity data storage (memory). This allows information about a part program to be entered (manually or from tape) and stored for future use.

Therefore, whenever the part program is required, it can be recalled from memory rather than reread from the tape. This not only protects the quality of the tape, but the information is recalled much faster.

11. Canned or Fixed Cycles

Storage capacity is generally provided in the MCU for any cycle (machining, positioning, etc.) which is used or repeated in a program. When the program is being written and a previous cycle must be repeated, all the programmer has to do is to insert a code in the program where that cycle is required. The control unit will recognize that code and recall from memory all the information required to perform that cycle (operation) again.

12. Subroutines and Macros

A parametric subroutine, sometimes called a "program within a program", is used to store frequently used data sequences (one block or a number of blocks of information) which can be recalled from memory as required by a code in the main part program.[2] An example of a subroutine could be a drilling cycle in which a series of 3/8 inch-diameter (9.5mm) holes 1 in. (25.4mm) deep must be drilled in a number of locations on a workpiece.

A macro is a group of instructions or data which are permanently stored and can be recalled as a group to solve recurring problems such as bolt-hole circle locations, drilling and tapping cycles, and other frequently used routines.[3] An example of a macro would be the XY locations of various holes on any bolt-circle diameter. When the diameter of the bolt circle and the number of holes on the circle are provided, the MCU makes all the calculations for hole locations and causes the machine tool slides to move into the proper position for each hole.

13. Axis Inversion

Axis inversion is the ability of the MCU to reverse plus and minus (+ and −) values along an axis or program zero to produce an accurate left-hand part from a right-hand program. This ability (called symmetrical machining) applies to all four quadrants and greatly reduces the time that would be required to program each part.

14. Digitizing

The digitizing feature allows a part program to be made directly from an existing part. The original part is traced while the CNC unit records all machine motions and produces a part program on punched tape.

15. Diagnostics

Diagnostic capabilities can monitor all conditions and functions of an NC machine or the control unit. If an error or malfunction occurs or if a changing condition nears a critical point, a signal or message is shown on the CRT. On some control units, as a critical point gets close, there may be a warning to the operator, shown on the screen, or the machine may automatically shut down. Diagnostic subroutines routines can be used to check hardware modules, circuit boards, and every area of the control unit to check their accuracy. The information is then displayed on the CRT.

16. Management Data

The modern MCU controls almost all machine tool functions through the build-in computer.

Since this information or data is already in the MCU controller or computer, it can be sent to the host or mainframe computer to provide valuable data to management and the operator. Spindle-on time, part-run time, number of parts machined, etc, can be recorded and sent to the host computer or displayed on the CRT screen.

Technical Words

computerize [kəm'pju:təraiz]	v. 用计算机处理；使计算机化
flexibility [ˌfleksə'biliti]	n. 灵活性，柔性
metric ['metrik]	adj. 米制的，公制的
interchange [ˌintə'tʃeindʒ]	v. 交换
parity ['pæriti]	n. 奇偶性
odd [ɔd]	adj. 奇数的，单数的
	n. 奇数
even ['i:vən]	adj. 偶数的
	n. 偶数
flaw [flɔ:]	n. 裂纹；缺陷
	v. 使破裂；使无效
subroutine [ˌsʌbru:'ti:n]	n. 子程序
macro ['mækrəu]	n. 宏指令，宏程序
	adj. 巨大的
inversion [in'və:ʃən]	n. 倒置
reverse [ri'və:s]	n. 相反
	adj. 相反的
	v. 颠倒
quadrant ['kwɔdrənt]	n. 象限
digitize ['didʒitaiz]	v. 将……数字化
diagnostics [ˌdaiəg'nɔstiks]	n. 诊断学，诊断技术
monitor ['mɔnitə]	n. 监视器
	v. 监控，监视
malfunction [mæl'fʌŋkʃən]	n. 故障
module ['mɔdju:l]	n. 模块

Technical Phrases

executive program	可执行程序
large-scale IC	大规模集成电路
cutting tool	切削刀具
American Standard Code for Information Interchange (ASCII)	
	美国信息交换标准代码
parity check	奇偶校验

circular interpolation	圆弧插补
linear interpolation	直线插补
parabolic interpolation	抛物线插补
helical interpolation	螺旋线插补
cubic interpolation	立方体插补
drilling cycle	钻孔循环
tapping cycle	加工螺纹循环
mirror image	镜像

Notes

1. The control unit automatically **calculates** what adjustments are necessary for the differences in size and **moves** the necessary slides to adjust for the difference.

句中有两个并列谓语动词：calculate 与 move；what adjustments are necessary for the difference in size 作 calculate 的宾语从句。

2. A parametric subroutine, sometimes called a "program within a program", is used to store frequently used data sequences (one block or a number of blocks of information) *which can be recalled from memory as required by a code in the main part program.*

句中 sometimes called a "program within a program" 是插入语；which 引导的定语从句作 data sequences 的定语。

3. A macro is a group of instructions or data which are permanently stored and can be recalled as a group to solve recurring problems such as bolt-hole circle locations, drilling and tapping cycles, and other frequently used routines.

宏指令是一组能够永久保存的指令或数据，可成组调用以解决重复出现的问题，如螺纹孔定位、钻孔、攻螺纹及其他频繁使用的子程序。

Exercises

(1) Place a " T " after sentences that are true and an " F " after those that are false.

1) It is usually the user not the manufacturer who changes the functions of a soft-wired or CNC unit.

2) Either a switch on the control unit or a specific code on the part program (G71 for inch and G70 for metric) will determine the measurement system used when machining a part.

3) The control unit identifies each code by the parity check: odd for the EIA code and even for the ASCII code.

4) Whenever the part program is required, it can be recalled from memory rather than reread from the tape in a CNC machine.

(2) Fill in the blanks according to the text with the words given below. Make changes if necessary.

   absolute   quadrant   breakthrough   malfunction   incremental

1) The development of a CNC system was made economically possible by electronic _____

which resulted in lower costs for minicomputers.

2) By using the proper address code G90 _____ or G91 _____, most CNC units will automatically program in that particular mode.

3) The ability of symmetrical machining applies to all four _____ and greatly reduces the time of programming each part.

4) If an error or _____ occurs or if a changing condition nears a critical point, a signal or message is shown on the CRT.

(3) Answer the following questions:

1) How many features does a CNC machine control unit of today have? What are they?

2) How many kinds of interpolation do the newer models include? What are they?

3) What are subroutines and macros? Please give us an example of each one.

【参考译文】

第10课 计算机数字控制

1970年，人们在数控词汇中引入了一个新术语——计算机数字控制（CNC）。电子技术的突破性发展降低了小型计算机成本，从而使CNC系统的开发经济可行。

无论是什么类型的数控机床，CNC机床是软件控制设备，它们的硬件组成都是相同的。制造商装入CNC计算机存储器的不是MCU，而是可执行程序或加载纸带。正是这个可执行程序才使控制装置像车削中心或加工中心一样具有"思考"功能。因此，适当改变可执行程序，即可改变CNC的功能。但改变可执行程序通常由制造商完成，无需用户考虑。

大规模集成电路与微处理器的运用使MCU已取得很大的发展，实现了几年前无法实现的许多功能，使NC系统柔性更强、效率更高。随着控制装置的不断发展，许多新功能会使机床加工性能更强。

与20世纪70年代前的硬件控制器相比，现代CNC机床（见图10-1）具有许多新功能：

1. 阴极射线管（CRT）

CRT就像一个电视屏幕，它有许多用途：

1) 显示加工过程中工作台或刀具的确切位置。

2) 显示整个加工程序，供编辑、修改时使用。

3) 协助安装工件。有些型号的CRT能接收传感器输送的信息，进而判断机床或控制器故障。

2. 绝对编程、增量编程

在选取G90（绝对方式）与G91（相对方式）方式之后，大多数CNC机床按已选代码模式自动编程。有些新式CNC机床还能在指定数据块中处理混合编程数据（绝对与增量）。

3. 英制、公制

大多数CNC机床既可使用英制，亦可使用公制。加工时既可用控制器上的开关，也可在程序中设定代码（G70表示英制、G71表示公制）选取要用的度量制。

4. EIA 代码、ASCII 代码

许多新式 CNC 机床能够识读美国电子工业学会（EIA）标准代码和美国信息交换标准（ASCII）代码。控制装置通过奇偶校验识别每一种代码：奇数为 EIA 代码，偶数为 ASCII 代码。

5. 手动数据输入

多数 CNC 机床能提供某种方法在必要时修改程序。这一功能很有必要，因为经常要根据工件规格修改程序、纠正程序错误、改变加工工序。

6. 程序编辑

几乎没有程序一开始就能免于出错，程序缺陷常常在加工时才会暴露出来。程序编辑功能有助于人们在控制器上直接校验、修改程序。

7. 插补

早期的控制装置仅能进行直线插补、圆弧插补、抛物线插补，而新式控制装置还能进行螺旋线与立方体插补。

8. 点定位与连续路径定位

所有 MCU 都能进行点定位、连续路径定位及组合形式定位。

9. 刀具直径补偿与长度补偿

在新式 MCU 中，可以手动输入刀具直径与刀具长度，并能根据不同程序来改变它们的值。控制器自动算出不同尺寸偏差需要的调整值，并使滑台作必要运动来补偿偏差。

10. 程序存储

新式 CNC 机床通常可提供大容量数据存储器，以便存储（手动或用纸带）零件程序信息，以备将来使用。因此，不论何时需要使用零件程序，都能从存储器中直接调用程序，而不是从纸带上重新读取程序。这样不仅保护了纸带质量，而且能更迅速地提取信息。

11. 固定循环

程序中常有重复使用的各种循环（如机床加工、定位等），MCU 通常都能提供空间存放循环程序。编写程序时，若要重复前面某个加工循环，编程员只需在程序适当地方插入循环调用代码，控制器会识别出这个代码，并从存储器中调用所有需要的信息，以再次执行该循环（操作）。

12. 子程序、宏指令

参数化子程序就是"程序中的程序"，专用于存放频繁调用的数据序列（指一个或几个信息块），需要时由主程序中的代码从存储器中直接调用。孔加工就是子程序的一个应用实例。如调用子程序可在工件几个不同位置上加工出直径 3/8 英寸（9.5mm）、深 1 英寸（25.4mm）的一系列孔。

宏指令是一组能够永久保存的指令或数据，可成组调用以解决重复出现的问题，如螺纹孔定位、钻孔、攻螺纹及其他频繁使用的子程序。宏指令的应用实例是确定螺纹直径系列上各孔 X、Y 坐标。当给定螺纹直径与圆周上孔的数目时，MCU 就会算出所有孔的位置坐标，并使机床滑台移动到各孔的准确位置。

13. 轴反演

轴反演是指 MCU 把右侧程序沿某轴或编程原点将正负颠倒，以产生精确的左侧部分。这一对称加工对四个象限都可适用，从而大大缩短了编制每个零件程序需要的时间。

14. 数字化功能

数字化功能可从实际零件直接生成加工程序。通过对原始零件的跟踪仿形，CNC 机床记录下所有的机床运动，并在穿孔带上生成加工程序。

15. 诊断功能

诊断功能可监视数控机床与控制器的所有运行状况与操作情况。若发生错误、运行异常，或加工状况接近临界点，CRT 屏幕会显示相应信号或信息。有些控制器上，当工况接近临界点时，会在屏幕上向操作者显示报警，或自动关闭机床。也可用诊断子程序检测硬件模块、电路板及控制器每个区域的精度，然后将相关信息显示于 CRT 屏幕。

16. 数据管理功能

现代 MCU 的内置计算机几乎控制了机床的所有功能。由于这种信息已存于 MCU 控制器或计算机中，因而可传送到主机，为管理者、操作者提供有价值的数据。主轴运行时间、零件加工时间、工件数量等都能记录下来，并送至主机，或显示于 CRT 屏幕。

Unit 11 Feeds & Speeds

Text

Feed

Feed is the amount that a cutting tool advances into the work, which generally controls the amount and rate that metal is removed from a workpiece. [1] It is generally measured in inches per revolution (in./r) or in inches per minute (in./min). On most NC machine tools, the feed rate is coded into the machine tool in inches per minute (in./min) by the manufacturer. The feed rate used will depend on the rigidity of the machine tool, the work setup, and the type of material being machined.

The EIA standard feed rate code consists of the letter F plus five digits (three to the left and two to the right of the decimal point) [2]. The numbers to the left of the decimal point represent whole inches (or millimeters), while the numbers to the right of the decimal represent fractions of an inch (or millimeter). Generally when a feed rate in inches or millimeters per minute is programmed, it is given in whole numbers (to the left of the decimal). A feed rate per revolution is generally given in decimals (to the right of the decimal point). Feed can be programmed in inches (or millimeters) per minute (F25.5 would be a feed rate of 25.5in./min, F0.01 would be a feed rate of 0.010in./r). The control unit on each machine tool will govern whether feed is programmed in inches perminute or inches per revolution. [3]

Speed

The speed of a machine tool spindle generally means the number of revolutions that the spindle makes in one minute of operation. The spindle speed rate (r/min) is generally governed by the work or cutter diameter, the type of cutting tool used, and the type of material being cut. Too fast a speed rate will cause the cutting tool to break down quickly, resulting in time wasted replacing or sharpening the cutting tool. Too slow a speed rate will result in the loss of valuable time, resulting in a higher cost for each part machined. Therefore the speed rate is a very important factor which affects the production rate and also the life of the cutting tool.

On NC machine control units, various methods are used to set the spindle speed. The most common are by revolutions per minute (r/min), surface feet (or meters) per minute (sf/min or sm/min) by the G96 function code which provides constant surface speed (CSS), or a three-digit code number ("Magic three").

The EIA recommends that spindle speed be programmed in revolutions per minute (r/min) [4]. The letter address S indicates spindle speed and may be followed by up to four digits. A spindle speed of 300r/min would be programmed as S300; a speed of 2,100 r/min would be programmed as S2,100.

Spindle speed may be programmed in surface feet (or meters) per minute through the G96 preparatory function code. Some MCUs (especially on turning centers) have the capabilities of maintaining CSS at the point of the cutting tool. The proper value of the surface speed is programmed under an appropriate G function code, and as a diameter changes during a machining operation, the spindle speed will automatically increase, decrease, or remain unchanged.

On some machine tools, spindle speeds can be programmed by using a three-digit coded number system, commonly called the "Magic Three". The second and third number of the three-digit code is the speed rounded off to two-digit accuracy. The first digit in the code always has a place value of 3 plus the number of digits to the left of the decimal point. For example, a speed of S610 would be as follows:

    6   means 000.000 (three numbers to the left and three numbers to the right of the decimal point)

    10 means 100.000   (100r/min)

The digits are always preceded by the letter address S; see the following example standard spindle speeds.

Code	r/min	Code	r/min
S570	70	S643	430
S610	100	S649	490
S614	140	S669	690
S615	150	S698	980
S620	200	S675	750
S622	220	S711	1100
S630	300	S715	1500
S635	350	S721	2100

Technical Words

revolution [ˌrevəˈluːʃən]	n. 旋转；转（数）
measure [ˈmeʒə]	n. 量度器；测量
	v. 测量；调节
rigidity [riˈdʒiditi]	n. 刚性；硬度
sharpen [ˈʃɑːpən]	v. 削尖，磨快，使尖锐
constant [ˈkɔnstənt]	n. 常数，恒量
	adj. 不变的；持续的

Technical Phrases

inch (es) per revolution (in/r)	英寸/转（in/r）
inch (es) per minute (in/min)	英寸/分钟（in/min）
feed rate	进给速度
speed rate	速率
decimal point	小数点
constant surface speed (CSS)	恒线切削速度

Notes

(1) Feed is the amount *that a cutting tool advances into the work*, which generally controls the amount and rate *that metal is removed from a workpiece.*

定语从句 that a cutting tool advances into the work 修饰先行词 the amount；which 引导的非限定性定语从句修饰主语 feed，该从句中 that 引导的定语从句修饰 the amount and rate。

(2) The EIA standard feed rate code consists of the letter F plus five digits (three to the left and two to the right of the decimal point).

短语动词 consist of 一般用主动形式，意为"由……组成"。

(3) The control unit on each machine tool will govern whether feed is programmed in inches per minute or inches per revolution.

govern（决定）后跟有一个宾语从句；whether... or... 意为"是……还是……"。

(4) The EIA recommends that spindle speed be programmed in revolutions per minute (r/min).

主句谓语动词 recommend 表"建议"，其后宾语从句要用虚拟语气，形式为：主语 + (should) + 动词原型，should 常省略。

Exercises

(1) Fill in the blanks according to the text with the words given below. Make changes if necessary.

　　　revolution　measure　rigidity　control unit

1) Feed is generally _____ in inches per revolution or in inches per minute.
2) One of the factors the feed rate depends upon is the _____ of the machine tool.
3) Whether feed is programmed in in./min or in./r is governed by the _____ on each machine tool.
4) Speed usually means the number of _____ a spindle makes in one minute of operation.

(2) Answer the following questions:

1) List the three factors which affect the feed rate used.
2) List the three factors which generally determine the speed of a machine tool spindle.
3) What consequences would result if too fast a speed is selected? What would result if too slow a speed?
4) What are the most common methods used to set the spindle speed?

【参考译文】

第 11 课　进给与转速

进给

进给是指刀具向工件前进的量，它通常影响金属切削量与切削速度，一般用英寸/转

(in/r)或英寸/分钟（in/min）作为单位。多数 NC 机床的进给已由制造厂家以英寸/分钟（in/min）作为单位编码并储存在机床中。至于加工中选用多大进给率，应视机床刚度、工件装夹以及待加工材料的类型进行选取。

EIA 标准进给代码由字母 F 加上 5 位数字（小数点左边三位、右边两位）组成。小数点左边的数字代表整数，其右边数字是小数部分，单位用英寸（或毫米）。以英寸/分钟、毫米/分钟为单位对进给编程时，一般给出整数部分（小数点左边部分），而每一转的进给常以小数给出（小数点右边的数字）。进给可用英寸/分钟（或毫米/分钟）编程（例如，F25.5 表示进给率是 25.5in/min，F0.01 表示进给率为 0.010in/r）。至于编程时是以英寸/分钟还是英寸/转作为进给单位，这是由每台机床的控制单元决定的。

转速

主轴转速通常是指机床运行一分钟主轴转过的周数。主轴速率（r/min）一般由工件或刀具直径、所用刀具的类型、工件材料的类型所决定。转速过高会很快崩断刀具，而更换、修磨刀具要浪费时间。转速过低又会浪费宝贵工时，使每个工件的加工成本升高。可见，转速是影响生产效率与刀具寿命的一大要素。

在 NC 机床控制单元中设定主轴转速的方法有多种，其中最通用的有：按每分种转数设定，单位用 r/min；用 G96 功能指令设定恒线切削速度，单位用英尺/分钟（f/min）或米/分钟（m/min）；使用"幻三码"。

编程时，EIA 推荐使用转/分钟（r/min）作为主轴转速单位。代码中，字母 S 代表主轴转速，其后最多有四个数字。主轴转速为 300r/min，可编程为 S300；而转速 2100r/min 则编程为 S2100。

此外，也可用 G96 功能码设定表面速度（英尺/分钟、米/分钟）对主轴转速编程。有些控制单元（尤其在车削中心上）能使刀刃保持恒定表面切削速度。适当的 G 功能码能够设定合理的表面速度，且主轴转速能随加工过程中刀具直径的变化而自动升高、降低或保持不变。

有些机床还可用一种所谓的"幻三码"对主轴转速编程。幻三码由三位数组成，其中第二、三位数字是由四舍五入得到的两位精度的数字；其第一位数字的数值总是由位权值 3 加上小数点左边数字的个数得到。例如，转速是 S610 的含义如下：

    6   表示   000.000（小数点左边三位，右边三位）
    10   表示   100.000（即 100r/min）

数字前总要冠以地址标志字母 S。请看以下标准主轴转速的例子：

代码	转速（r/min）	代码	转速（r/min）
S570	70	S643	430
S610	100	S649	490
S614	140	S669	690
S615	150	S698	980
S620	200	S675	750
S622	220	S711	1100
S630	300	S715	1500
S635	350	S721	2100

Unit 12 Continuous Path

Text

Contouring, or continuous path machining, involves work such as that produced on a lathe or milling machine, where the cutting tool is in contact with the workpiece as it travels from one programmed point to the next.[1] Continuous path positioning is the ability to control motions on two or more machine axes simultaneously to keep a constant cutter-workpiece relationship. The programmed information on the NC tape must accurately position the cutting tool from one point to the next and follow a predefined accurate path at a programmed feed rate in order to produce the form or contour required.

The method by which contouring machine tools move from one programmed point to the next is called interpolation. This ability to merge individual axis points into a predefined tool path is built into most of today's MCUs.[2] There are five methods of interpolation: linear, circular, helical, parabolic, and cubic. All contouring controls provide linear interpolation, and most controls are capable of both linear and circular interpolation. Helical, parabolic, and cubic interpolation are used by industries that manufacture parts which have complex shapes, such as aerospace parts and dies for car bodies.

Linear Interpolation

Linear interpolation consists of any programmed points linked together by straight lines, whether the points are close together or far apart. Curves can be produced with linear interpolation by breaking them into short, straight-line segments. This method has limitations, because a very large number of points would have to be programmed to describe the curve in order to produce a contour shape.

A contour programmed in linear interpolation requires the coordinate positions (XY positions in two-axis work) for the start and finish of each line or segment. Therefore, the end point of one line or segment becomes the start point for the next segment, and so on, throughout the entire program.

The accuracy of a circle or contour shape depends on the distance between each two programmed points. If the programmed points (XY) are very close together, an accurate form will be produced. To understand how a circle can be produced by linear interpolation, refer to Fig. 12-1a. Fig. 12-1b shows that when 8 connecting lines are machined, an octagon is produced, while Fig. 12-1c shows 16 connecting or chord lines. If the chords in each are examined, it should be apparent that the more program points there are, the more closely the form resembles a circle.[3] If there were 120 chord lines, it would be difficult to see that the circle was made up of a series of short chord lines. Therefore, on a control unit which is capable only of linear interpolation, very accurate contours require very long programs because of the large number of points which need to be programmed.

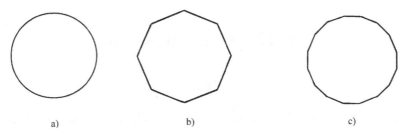

Fig. 12-1   Linear interpolation with arcs and circles
a) True circle   b) 8-Segment "circle"   c) 16-Segment "circle"

Circular Interpolation

The development of MCUs capable of circular interpolation has greatly simplified the process of programming arcs and circles. To program an arc, the MCU requires only the coordinate positions (the *XY* axes) of the circle center, the radius of the circle, the start point and end point of the arc being cut, and the direction in which the arc is to be cut (clockwise or counterclockwise) (Fig. 12-2). The information required may vary with different MCUs.

The circular interpolator in the MCU breaks up the distance of each chord line (circular span) into a series of the smallest movement increments caused by one single output pulse. This is usually 0.0001 in. (0.002mm), and the interpolator automatically computes enough output pulses to describe the circular form and then provides the control data required for the cutting tool to produce the form.

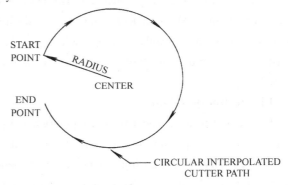

Fig. 12-2   Two-dimensional circular interpolation

The advantage and power of circular interpolation are better appreciated by comparing it to linear interpolation. More than a thousand blocks of information would be needed to produce a circle in linear interpolation segment by segment.[4] Circular interpolation requires only five blocks to produce the same circle.

Helical Interpolation

Helical interpolation combines two-axis circular interpolation with a simultaneous linear movement in the third axis. All three axes move at the same time to produce the helical (spiral) path required. Helical interpolation is most commonly used in milling large internal-diameter threads or helical forms.

Parabolic Interpolation

Parabolic interpolation has its greatest application in automotive dies, mold work, and any form of sculpturing. Parabolic interpolation is a method of creating a cutter path covering a wide variety of geometric shapes, such as circles, ellipses, parabolas, and hyperbolas. Parabolic interpolation can be defined as a movement that is either totally parabolic or part parabolic and has three non-straight-line locations (two end points and one midpoint). This allows it to closely simulate curved sections, using about 50 times fewer program points than would be required using linear interpolation. The more powerful computers of today have put parabolic interpolation in the background because it is now possible to use simpler interpolation.

Cubic Interpolation

Cubic interpolation enables sophisticated cutter paths to be generated in order to cut the shapes of the forming dies used in the automobile industry. These dies can be machined with only a small number of input data points in the program. Besides describing the geometry, cubic interpolation smoothly blends one curved segment to the next without the interruption of boundary (start and stop) points. However, a larger-than-normal computer memory is required to handle the extensive program for cubic interpolation.

Technical Words

contour [ˈkɔntuə]	n.	轮廓，轮廓线，外形
	v.	加工轮廓
predefine [ˌpriːdiˈfain]	v.	预先确定
merge [məːdʒ]	v.	合并，结合
curve [kəːv]	n.	曲线
	v.	使弯曲
octagon [ˈɔktəgən]	n.	八边形，八角形
chord [kɔːd]	n.	弦
simplify [ˈsimplifai]	v.	简化，单一化
clockwise [ˈklɔkwaiz]	adj.	顺时针的
counterclockwise [ˌkauntəˈklɔkwaiz]	adj.	逆时针的
spiral [ˈspaiərəl]	adj.	螺旋形的
	n.	螺旋
thread [θred]	n.	螺纹
ellipse [iˈlips]	n.	椭圆，椭圆形
parabola [pəˈræbələ]	n.	抛物线
hyperbola [haiˈpəːbələ]	n.	双曲线
simulate [ˈsimjuleit]	v.	模拟，模仿

Technical Phrases

continuous path positioning	连续路径定位
straight-line segment	直线段
automotive die	汽车冲模
geometric shape	几何形状
forming die	成型模
automobile industry	汽车工业

curved surface   曲面

Notes

(1) Contouring, or continuous path machining, involves work such as that produced on a lathe or milling machine, where the cutting tool is in contact with the workpiece as it travels from one programmed point to the next.

轮廓加工又称为连续路径加工,它是在车床或铣床上进行的,当刀具从一个编程点移动到另一个编程点时,刀具与工件始终保持接触。

(2) This ability to merge individual axis points into a predefined tool path is built into most of today's MCUs.

不定式 to merge… path 作定语修饰 ability; merge… into… 意为"把……合并为……"。

(3) If the chords in each are examined, it should be apparent that the more program points there are, the more closely the form resembles a circle.

the more…, the more… 的形式意为"越……越……"。

(4) More than a thousand blocks of information would be needed to produce a circle in linear interpolation segment by segment.

若用直线插补一段一段地生成圆,需要一千多段程序信息。

Exercises

(1) Place a "T" after sentences that are true and an "F" after those that are false.

1) Linear interpolation consists of any programmed points joined together by straight lines.

2) The circular interpolator in the MCU breaks up the distance of each chord line into a series of the smallest movement increments caused by one single output pulse.

3) A normal computer memory is required to handle the extensive program for cubic interpolation.

4) Circular interpolation uses more blocks to produce a circle than linear interpolation.

(2) Fill in the blanks according to the text with the words given below. Make changes if necessary.

  simplify segment parabolic chord interpolation

1) The method by which contouring machine tools move from one programmed point to the next is called _____.

2) A contour programmed in linear interpolation requires the coordinate positions (*XY* positions in two-axis work) for the start and finish of each line or _____.

3) If the _____ in each are examined, it should be apparent that the more program points there are, the more closely the form resembles a circle.

4) The development of MCUs capable of circular interpolation has greatly _____ the process of programming arcs and circles.

5) _____ interpolation is a method of creating a cutter path covering a wide variety of geo-

metric shapes, such as circles, ellipses, parabolas, and hyperbolas.

(3) Answer the following questions:
1) What is continuous path positioning?
2) How many methods of interpolation are mentioned in the text, and what are they?
3) What is the function of interpolation and the purpose of nonlinear interpolation?

【参考译文】

第12课 连续路径加工

轮廓加工又称为连续路径加工，它包括像在车床或铣床上那样进行的零件加工，当刀具从一个编程点移动到另一个编程点时，刀具与工件始终保持接触。连续路径定位是指控制两轴或多轴联动以保持恒定的刀具——工件相对位置的能力。NC纸带上的编程信息能够使刀具从一点到另一点准确定位，并按程序设定的进给速度沿预定路径运动，从而加工出所要求的形状或轮廓。

轮廓加工时，刀具从一个编程点移动到下一个相邻编程点采用的方式称为插补。现代MCU大多都有将各个轴向的点复合为预定走刀路径的功能。插补方法有五种：直线插补、圆弧插补、螺旋线插补、抛物线插补、立方体插补。所有轮廓加工机床都有直线插补功能，多数机床还兼有直线与圆弧两种插补功能。在制造业中，螺旋线插补、抛物线插补、立方体插补可加工各种复杂型面的零件，例如航空器零件与汽车车身模具。

直线插补

直线插补用线段把许多相距或近或远的编程点连起来。若把曲线分成许多短直线段，则可用直线插补加工该曲线。但是，直线插补有局限性，因为要生成某一轮廓形状，必须对大量的点进行编程才能描述曲线形状。

用直线插补进行轮廓编程时，需要用每条线段的起点与终点坐标（即平面工件的 XY 坐标）。因此，在整个程序中，一条线段的终点就是相邻下一条线段的起点。

直线插补中，圆弧或轮廓的精度取决于每两个相邻编程点之间的距离。如果编程点（XY 坐标）彼此非常靠近，加工出的形状就准确一些。为了理解如何运用直线插补加工圆形，可参阅图12-1a。图12-1b表明，如果加工8条首尾连接的线段，则形成一个8边形。图12-1c表示了加工16段连接弦线时的情形。如果检查每个图中弦线数目，显然编程点越多，加工出的形状就越接近于圆。假如用120段弦线，就很难看出该圆竟是由许多很短的弦线组成。因此，对只能直线插补的控制单元，要想加工精确轮廓，就需要很长的程序，因为编程点数量庞大。

圆弧插补

具有圆弧插补功能的MCU的出现极大地简化了对圆与圆弧的编程。为了对圆弧编程，MCU只需圆心坐标（XY 轴）、圆弧半径、圆弧起点与终点坐标、圆弧方向（顺时针、逆时针）（见图12-2）。对于不同的MCU，圆弧插补需要的参数也就不同。

MCU的圆弧插补器把每段弦线（或圆弧跨距）分成许多由输出单个脉冲引起的最小移动增量，其值通常为0.0001英寸（0.002mm）。插补器自动算出所需脉冲个数，然后形成加工这个形状时需要的刀具控制数据。

若与直线插补相比，圆弧插补的优点与能力备受赞赏。若用直线插补一段一段地生成圆，需要一千多段程序信息，而用圆弧插补加工出同样圆形却只需要五段程序。

螺旋线插补

螺旋线插补可将两轴圆弧插补与同时沿第三轴向的直线运动复合起来。由于三轴同时运动，因而可按需要形成螺旋状路径。螺旋线插补多用于大内径螺纹或螺旋形状的铣削加工。

抛物线插补

抛物线插补在汽车冲模、铸模以及其他雕刻形状的加工中应用最广。抛物线插补是一种能生成多种不同几何形状加工路径的方法，如圆、椭圆、抛物线及双曲线。抛物线插补有完全抛物线与部分抛物线之分，需要三个非共线定位点（两个端点、一个中点）。抛物线插补可非常逼真地模拟曲线段，而所用的程序段比用直线插补的程序段少50倍。因为目前可以使用更简单的插补功能，现代功能强大的计算机已把抛物线插补作为备选功能。

立方体插补

立方体插补能生成复杂形状的走刀路径，以便加工汽车工业用到的成型模具形状。立方体插补仅用少量编程点的数据就可加工出这些模具。除了描述几何图形外，立方体插补能够把一段曲线与下一段曲线光滑连接而没有边界中断点（起点和终点）。然而，立方体插补需要用超大计算机内存，以便处理立方体插补庞大的程序。

Unit 13 Types and Parts of Machining Centers

Text

There are two main types of machining centers: the horizontal spindle machine and the vertical spindle machine.

Types of Machining Centers:

1. Horizontal Spindle Type

(1) The traveling-column type (Fig. 13-1) is equipped with one or usually two tables on which the work can be mounted. With this type of machining center, the workpiece can be machined while the operator is loading a new workpiece on the other table.

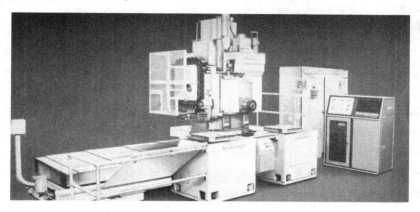

Fig. 13-1 A traveling-column machining center

(2) The fixed-column type (Fig. 13-2) is equipped with a pallet shuttle. The pallet is a removable table on which the workpiece is mounted. After the workpiece has been machined, the workpiece and pallet are moved to a shuttle which then rotates, bringing a new pallet and workpiece into position for machining. [1]

2. Vertical Spindle Type

The vertical spindle machining center (Fig. 13-3) is a saddle-type construction with sliding bedways which utilizes a sliding vertical head instead of a quill movement.

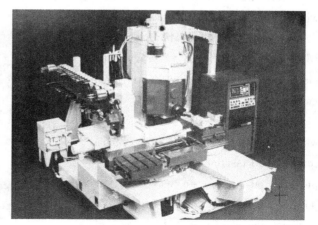

Fig. 13-2 A fixed-column machining center

Parts of the CNC Machining Centers

The main parts of a CNC machining center are the bed, saddle, column, table, servo system, spindle, tool changer, and the machine control unit (MCU) (Fig. 13-4).

Fig. 13-3 A vertical spindle machining center

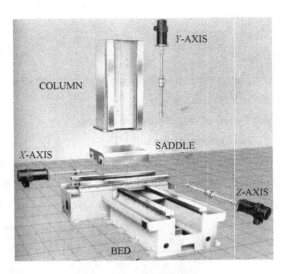

Fig. 13-4 The main parts of a CNC machining center

Bed—The bed is usually made of high-quality cast iron which provides for a rigid machine capable of performing heavy-duty machining and maintaining high precision.[2] Hardened and ground ways are mounted to the bed to provide rigid support for all linear axes.

Saddle—The saddle, which is mounted on the hardened and ground bedways, provides the machining center with the X-axis linear movement.

Column—The column, which is mounted to the saddle, is designed with high torsional strength to prevent distortion and deflection during machining. The column provides the machining center with the Y-axis linear movement.

Table—The table, which is mounted on the bed, (Fig. 13-5) provides the machining center with the Z-axis linear movement.

Servo system—The servo system, which consists of servo drive motors, ball screws, and position feedback encoders, provides fast, accurate movement and positioning of the XYZ axes slides. The feedback encoders mounted on the ends of the ball screws form a closed-loop system which maintains consistent high-positioning unidirection repeatability of ± 0.0001 in. (0.00254mm).[3]

Fig. 13-5 The table provides the machining center with the Z-axis linear movement

Spindle—The spindle, which is programmable in 1r/min increments, has a speed range of from

20 to 6000 r/min. The spindle can be of a fixed position (horizontal) type, or can be a tilting/contouring spindle which provides for an additional A axis (Fig. 13-6).[4]

Tool changers—There are basically two types of tool changers, the vertical tool changer (Fig. 13-7) and the horizontal tool changer (Fig. 13-8). The tool changer is capable of storing a number of preset tools which can be automatically called for use by the part program. Tool changers are usually bi-directional, which allows for the shortest travel distance to randomly access a tool. The actual tool change time is usually only 3 to 5s.

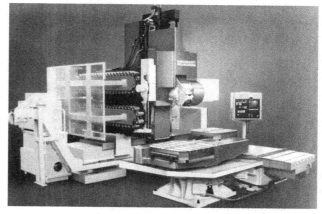

Fig. 13-6 The tilting contouring spindle provides an additional A axis

MCU—The MCU allows the operator to perform a variety of operations such as programming, machining, diagnostics, tool and machine monitoring, etc. MCUs vary according to manufacturers' specifications; new MCUs are becoming more sophisticated, making machine tools more reliable and the entire machining operations less dependent on human skills.

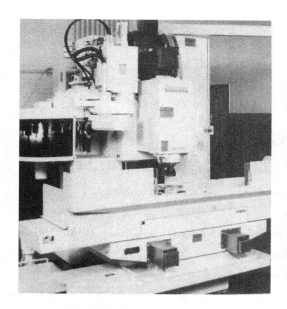

Fig. 13-7 Vertical tool changer

Fig. 13-8 Horizontal tool changer

Technical Words

equip [iˈkwip]	v.	装备，配备
mount [maunt]	n.	装配

		v. 安装；设置；安放
pallet ['pælit]		n. 平板架
shuttle ['ʃʌtl]		n. 滑闸，滑台
		v. 穿梭往返
saddle ['sædl]		n. 鞍，鞍状物
bedway ['bedwei]		n. 床身导轨
quill [kwil]		n. 衬套；主轴
harden [hɑːdn]		v. 使变硬；淬火
strength [streŋθ]		n. 力；强度
distortion [dis'tɔːʃən]		n. 扭曲；变形；失真
deflection [di'flekʃən]		n. 偏斜；偏转；偏差
encoder [in'kəudə]		n. 编码器
increment ['inkrimənt]		n. 增加；增量
bidirectional [ˌbaidi'rekʃənəl]		adj. 双向的

Technical Phrases

traveling-column	移动立柱
fixed-column	固定立柱
ball screw	滚珠丝杠
tool changer	换刀机构
machine control unit (MCU)	机床控制单元
cast iron	铸铁
torsional strength	扭转强度

Notes

(1) After the workpiece has been machined, the workpiece and pallet are moved to a shuttle *which then rotates*, bringing a new pallet and workpiece into position for machining.

定语从句 which then rotates 修饰 a shuttle。

(2) The bed is usually made of high-quality cast iron *which provides for a rigid machine capable of performing heavy-duty machining and maintaining high precision*.

which 引导的定语从句修饰 high-quality cast iron。

(3) **The feedback encoders** mounted on the ends of the ball screws **form a closed-loop system** which maintains consistent high-positioning unidirection repeatability of ± 0.0001in. (0.00254mm).

主句结构为：The feedback encoders… form a closed-loop system…。过去分词短语 mounted on the ends of the ball screws 做后置定语修饰 encoders；which 引导的定语从句修饰 a closed-loop system。

(4) The spindle can be of a fixed position (horizontal) type, or can be a tilting/contouring

spindle which provides for an additional *A* axis.

主句谓语采用 can be of 的形式，be of 做谓语常表示事物的属性。

Exercises

（1）Fill in the blanks according to the text with the words or phrases given below. Make changes if necessary.

  cast iron access tool changer movement ball screw

1）One main part of the CNC machine is the bed which is made of high-quality _____.

2）Saddle provides a machining center with the *X*-axis linear _____.

3）A servo system consists of servo motors, _____, and position feedback encoders.

4）A _____ stores a number of preset tools which can be automatically called for use by a part program.

5）Tool changers are usually bidirectional, therefore a tool can be _____ in a random way with the shortest travel distance.

（2）Answer the following questions：

1）There are two main types of machining centers, and what are they?

2）What are the main parts of a CNC machining center?

3）With what type of machining center can a workpiece be machined while the operator is loading a new one on the other table?

4）List the two sorts of tool changers.

【参考译文】

第 13 课　加工中心的种类与组成

加工中心分为两大类：一是卧式主轴加工中心，二是立式主轴加工中心。

加工中心的种类：

1. 卧式主轴加工中心

（1）移动立柱加工中心（见图 13-1）通常配有一或两个工作台来装夹工件。这种加工中心可一边加工工件，一边由操作人员在另一个工作台上装夹工件。

（2）固定立柱加工中心（见图 13-2）上配有一个平板架，这是一个装有工件的可移动工作台。完成加工后，工件与平板架一起移到滑台，滑台然后转动，把新的平板架与工件输送到加工位置。

2. 立式加工中心

立式加工中心（见图 13-3）呈鞍形结构，其中装有滑动导轨，该导轨采用立式滑块运动，而不是主轴运动。

CNC 加工中心的组成部件

CNC 加工中心的主要部件有：床身、床鞍、立柱、工作台、伺服系统、主轴、换刀机构、机床控制单元（MCU）（见图 13-4）。

床身——床身通常用优质铸铁制成，使机床具有刚性加工能力，可实现重型加工，并能保持良好的加工精度。经过淬火和磨削的平导轨装在床身上，为各直线轴提供刚性支撑。

床鞍——床鞍装在经过淬火和磨削的平导轨上，为加工中心提供 X 轴方向的直线运动。

立柱——立柱装在床鞍上，具有很高的扭曲强度，以防止加工过程中的变形与偏移。立柱为加工中心提供 Y 轴方向的直线运动。

工作台——工作台装于床身上（见图13-5），为加工中心提供 Z 轴方向的直线运动。

伺服系统——伺服系统由伺服驱动电动机、滚珠丝杠、位置反馈编码器组成，它为 XYZ 轴滑台提供快速准确运动与定位。反馈编码器安装在滚珠丝杠端部构成闭环系统，单向重复定位精度可达±0.0001英寸（0.00254mm）。

主轴——主轴可以以1r/min的增量编程，其转速在20～6000r/min之间。主轴可以是固定式（卧式），或是倾斜/轮廓主轴，以提供一个附加轴（A轴）（见图13-6）。

换刀机构——换刀机构有两种基本类型，即立式换刀机构（见图13-7）与卧式换刀机构（见图13-8）。换刀机构中存放了许多预选刀具，由加工程序自动调用。换刀机构一般是双向的，允许以最短路径随机存取刀具。实际换刀时间通常只有3～5s。

机床控制单元——机床控制单元允许操作员进行各种操作，例如编程、加工、诊断、刀具与机床监控等。生产规格不同，控制单元也不同。新型控制单元采用尖端技术，使机床性能更可靠，并使整个加工过程对人为技能的依赖越来越少。

【Reading Material】

Position Detectors—Opto-Electronical Rotary Encoders

Opto-electronical rotary encoders are used for measuring displacement, angles of rotation or speeds on machines. They can be used in conjunction with numerical controls, programmable logic controllers, drives and position displays.

A basic differentiation is made between incremental and absolute measurement techniques. Because the position of the control is not normally stored, and movements of the machine are not recorded during an emergency stop, the use of incremental encoders requires that the machine carry out a reference point approach after every emergency stop.

In contrast, absolute value encoders also record these movements and supply the current position after emergency stops. Therefore, a reference point approach is not required.

The power supply for the encoder is 5VDC or optionally 10 to 30VDC. The 10 to 30V version permits longer cable lengths. Most control systems provide the supply voltage directly from the measuring circuit.

Incremental encoders

Incremental encoders supply a defined number of electrical pulses for each revolution. The number of revolutions constitutes a measurement of the traveled path or angle. Incremental encoders are operated by scanning code disks photo-electrically using the backlighting method. A light-emitting diode (LED) is used as the source. Photoelectric elements record the light-dark modulation caused by the turning encoder shaft. The suitable arrangement of the bar pattern on the coded disk attached to the shaft and the fixed aperture permits the photoelectric elements to supply A and B track signals that are displaced at 90° to each other and also a reference signal R.

The encoder amplifies these signals and converts them into various output levels.

Absolute value encoders

Absolute value encoders (angular encoders) have the same design as incremental encoders with regard to the sampling principle. But they possess a larger number of tracks up to 12 (tracks).

A single step code (gray code) is used. This avoids the occurrence of sampling errors. After the machine has been switched on, the position value is transferred immediately to the control. A reference point approach is not necessary. The data transmission between encoder and control is performed through the synchronous serial interface (SSI) or through PROFIBUS-DP.

Unit 14 Tool Monitoring and In-Process Gaging

Text

Tool Monitoring

When machining is down on a turning center, tool wear and breakage require the continuous attention of the operator. A tool-monitoring system can be substituted for the skilled operator's eyes and ears, signaling in a variety of ways the need to replace tools that are worn and broken. [1]

There are many types of tool-monitoring systems available, and the way they detect tool wear varies with the manufacturer. Given that a worn or dull tool requires more power to machine a workpiece than a sharp tool, the most common method of determining wear is from the power or force it takes to drive the cutting tool. The tool-monitoring system measures the load on the main spindle drive motor. This is done in two stages:

(1) When the machine is set up and ready to produce a workpiece, the normal machining cycle is run with new tools and all speeds and feeds at 100 percent of programmed rate.

(2) When the workpiece is completed, a second machining cycle is run, this time without contacting the workpiece.

From these two cycles the monitoring system can calculate the net machining forces and torque for every portion of the part program where monitoring is desired. [2] Once the limits have been set, the monitoring system will signal the operator when machining forces and torque exceed acceptable limits and, in some cases, automatically reduce the speed or feed to compensate for the dullness of the cutting tool.

Some other methods of detecting tool wear and breakage are by:
- Sound
- Electrical resistance
- Vibration
- Radioactivity
- Measuring heat
- Optical magnification

Regardless of the system of detection used, the tool-wear-monitoring system provides such benefits as:
- Broken tool detection
- Reduced operator attention
- Machine protection
- Worn tool detection
- Improved productivity

In-Process Gaging

An in-process gaging system is a way of monitoring what is happening to the workpiece and tools during the machining operations.[3] It can also be used to compensate for tool wear and thermal growth, determine tool offsets and locate workpieces, etc.

The probe works basically by sending a signal to the machine control as soon as the probe has been deflected in any direction upon making contact with the workpiece or the tool.[4] In-cycle gages are omnidirectional, which means that the sensing probes will sense any $\pm X$, $\pm Y$, or $\pm Z$ direction. Once the probe stylus has been deflected, a signal is automatically sent to the control where the data can be acted on.

On chucking and turning centers, the probes can be mounted either in a toolholder or in a turret. The probes can be selected the same way a cutting tool is selected—by calling it up in the machining program—to check a tool for wear and make the appropriate compensation or to check a part for size between machining operations.

The use of in-process gaging helps to reduce operator errors in setup and allows for inspection of fully machined parts in the machine.

Technical Words

wear [wɛə]	v.	(~ out) 磨损
detect [di'tekt]	v.	探测；检测
dull [dʌl]	v.	使变钝
torque [tɔːk]	n.	扭矩，转矩
compensate ['kɔmpənseit]	v.	补偿
breakage ['breikidʒ]	n.	破坏；破损
vibration [vai'breiʃən]	n.	振动，颤动
resistance [ri'zistəns]	n.	阻力；电阻
radioactivity [ˌreidiəuæk'tiviti]	n.	放射能，辐射性
optical ['ɔptikəl]	adj.	光学的
magnification [ˌmægnifi'keiʃən]	n.	扩大；放大倍数
datum ['deitəm]	n.	数据；资料
probe [prəub]	n.	探针，探测器；
	v.	（以探针等）探察
deflect [di'flekt]	v.	（使）偏斜；（使）偏转
omnidirectional [ˌɔmnidi'rekʃənl]	adj.	全方向的
stylus ['stailəs]	n.	测针，触头
chuck [tʃʌk]	n.	卡盘，夹具
	v.	用卡盘夹住
toolholder ['tuːlˌhəuldə(r)]	n.	刀柄，刀架

Technical Phrases

tool monitoring	刀具监测
in-process gaging	在程检测

tool offset　　　　　　　　　　刀具偏移量

Notes

(1) A tool-monitoring system can be substituted for the skilled operator's eyes and ears, signaling in a variety of ways the need to replace tools *that are worn and broken.*

in a variety of ways 意为"用各种方法"。定语从句 *that are worn and broken* 修饰 tools。

(2) From these two cycles the monitoring system can calculate the net machining forces and torque for every portion of the part program *where monitoring is desired.*

where 引导的定语从句修饰 the part program；net machining forces 指"净加工力"，net machining torque 指"净加工转矩"。

(3) An in-process gaging system is a way of monitoring what is happening to the workpiece and tools during the machining operations.

what 引导的是一个宾语从句。

(4) The probe works basically by sending a signal to the machine control as soon as the probe has been deflected in any direction upon making contact with the workpiece or the tool.

as soon as 意为"一……就……"。

Exercises

(1) Fill in the blanks according to the text with the words given below. Make changes if necessary.

　　torque　　sense　　monitor　　load　　detect

1) The way different tool-monitoring systems _____ tool wear varies with the manufacturer.

2) According to the second paragraph, the most common method of determining tool wear is to measure the _____ on the main spindle drive motor.

3) An in-process gaging system provides a way of _____ what is happening to the workpiece and tools during the machining operations.

4) With an in-cycle gage, any ±X, ±Y, or ±Z direction can be _____ by the sensing probes.

5) The tool monitoring system will send the operator a signal once machining _____ exceeds the preset limits.

(2) Translate the following sentences into Chinese:

1) Tool wear and breakage used to need continuous attention of the machine operator.

2) A worn tool needs more power to machine a workpiece than a sharp tool.

(3) Answer the following questions:

1) What is the most common method of determining tool wear?

2) When the tool-monitoring system measures the load on the main spindle drive motor, which two stages should be taken?

3) What benefits can a tool-wear monitoring system provide?

【参考译文】

第14课 刀具监控及在程检测

刀具监控

在车削中心上加工工件时,操作人员要时刻关注刀具的磨损与崩碎情况。刀具监控系统相当于一位熟练工的眼睛和耳朵,并以各种不同方式发出信号,提示更换已磨损或崩碎的刀具。

刀具监控设备的种类很多,生产厂家不同,检测刀具磨损的方式也就不同。鉴于使用磨钝刀具比锋利刀具加工所需能量要多一些,因此,驱动功率或驱动力就成为判断刀具磨损程度最常用的量。测量主轴驱动电动机上的负载,即可得到驱动功率的大小。这个过程通常分两步进行:

(1)调整机床、装好刀具后,使机床以100%的编程进给率全速运行一个加工周期。

(2)完成工件加工后,运行第二个加工周期,但刀具与工件互不接触。

根据这两个加工周期的运行情况,机床监控设备可算出需要监控的零件程序中的各部分所需的净加工力与净加工转矩。一旦设定了这些极限值,每当加工力或力矩超过极限时,监控设备就会向操作人员发出提醒信号;在某些情况下,还可自动降低转速或进给率,补偿刀具磨损。

除此以外,还可通过以下各量检测刀具的磨损情况:

- 声音
- 电阻
- 振动
- 辐射
- 热量
- 光电放大率

无论采用哪一种检测方法,刀具磨损检测设备都有以下优点:

- 检测崩损刀具
- 减少操作人员的疲劳程度
- 保护机床
- 检测磨损刀具
- 提高生产效率

在程检测

在程检测是指在加工过程中检测工件与刀具状况的方法。在程监测还可用于补偿刀具磨损及热膨胀、确定刀具偏移量、工件定位等。

探针的基本工作原理是:探针一旦与工件或刀具接触而沿某一方向偏转时,便立即给机床控制部分发出一个信号。在程检测是全方位的,其传感探针能够检测 $\pm X$、$\pm Y$、$\pm Z$ 的任一方向。探针的针尖一旦偏斜,便会向控制部分自动发出信号,由控制单元对数据进行处理。

对于卡盘与车削中心来说,探针可安装在刀架或转塔上。探头的选取方式与刀具选取方式相同,即在加工程序中直接调用。在程监测既可检测刀具磨损状况并进行适当补偿,亦可在加工期间检验零件尺寸。

运用在程检测可以减少操作人员调试机床时的失误,并能检验成品工件。

Unit 15   Cutting Tools

Text

The selection of the proper cutting tools for each operation on a machining center is essential to producing an accurate part. Generally there is not enough thought and planning going into the selection of cutting tools for each particular job. The NC programmer must have a thorough knowledge of cutting tools and their applications in order to properly program any part.

Machining centers use a variety of cutting tools to perform various machining operations. These tools may be conventional high-speed steel, cemented carbide inserts, CBN (cubic boron nitride) inserts, or polycrystalline diamond insert tools. Some of the tools used are end mills, drills, taps, reamers, boring tools, etc.

Studies show that machining center time consists of 20 percent milling, 10 percent boring, and 70 percent hole-making in an average machine cycle. On conventional milling machines, the cutting tool cuts approximately 20 percent of the time, while on machining centers the cutting time can be as high as 75 percent. The end result is that there is a larger consumption of disposable tools due to decreased tool life through increased tool usage.

End Mills

End mills (Fig. 15-1) and shell end mills are widely used in machining centers. They are capable of performing a variety of machining operations such as face, pocket and contour milling, spotfacing, counterboring, and roughing and finishing of holes using circular interpolation.

Drills

Conventional as well as special drills are used to produce holes (Fig. 15-2). Always choose the shortest drill that will produce a hole of the required depth. As drill diameter and

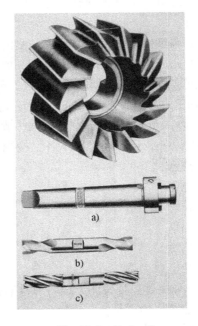

Fig. 15-1   End mills
a) Shell end mill and adapter   b) Two-flute end mill
c) Four-flute end mill

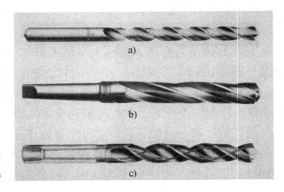

Fig. 15-2   Stub drills
a) A high helix drill   b) A core drill   c) An oil hole drill

length increase, so does the error in hole size and location.[1] Stub drills are recommended for drilling on machining centers.

Center Drills

Center drills (Fig. 15-3) are used to provide an accurate hole location for the drill which is to follow. The disadvantage of using center drills is that the small pilot drill can break easily unless care is used. An alternative to the center drill is the spotting tool, which has a 90° included angle and is widely used for spotting hole locations.

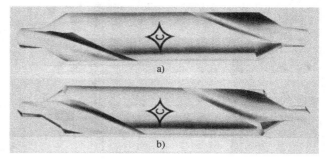

Fig. 15-3　Center drills
a) Regular type　b) Bell type

Taps

Machine taps (Fig. 15-4) are designed to withstand the torque required to thread a hole and clear the chips out of the hole. Tapping is one of the most difficult machining operations to perform because of the following factors:

● Inadequate chip clearance

● Inadequate supply of cutting fluid

● Coarse and fine threads in various materials

● Speed and feed of threading operations being governed by the lead of the thread

● Depth of thread required

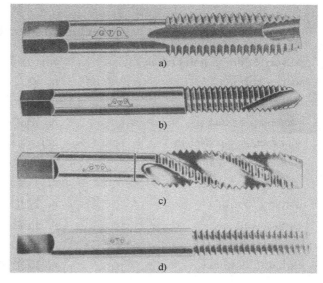

Fig. 15-4　Taps
a) Gun　b) Stub flute　c) Spiral flute　d) Fluteless

Reamers

Reamers (Fig. 15-5) are available in a variety of designs and sizes. A reamer is a rotary end cutting tool used to accurately size and produce a good surface finish in a hole which has been previously drilled or bored[2].

Boring Tools

Boring is the operation of enlarging a previously drilled, bored, or cored hole to an accurate size and location with a desired surface finish. This operation is generally performed with a single-point boring tool (Fig. 15-6). When a boring bar is selected, the length and diameter should be carefully considered: as the ratio between length and diameter increases, the rigidity of the boring bar decreases. For example, a boring bar with a 1:1 length-to-diameter ratio is 64 times more rigid than the one with 4:1 ratio.

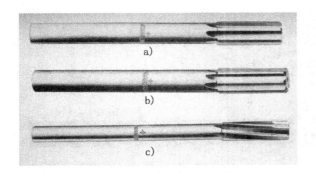

Fig. 15-5 Reamers
a) Rose reamer b) Fluted reamer c) Carbide-tipped reamer

Fig. 15-6 Single-point boring tools

Technical Words

polycrystalline [ˌpɔli'kristəlain]	adj. 多晶的
reamer ['ri:mə]	n. 铰刀，扩锥
mill [mil]	n. 铣床；铣刀
contour ['kɔntuə]	n. 轮廓
spot-facing [spɔt'feisiŋ]	adj. 刮孔口平面的
counterbore [ˌkauntə'bɔ:]	v. 反镗
tap [tæp]	n. 丝锥
boring ['bɔ:riŋ]	n. 镗孔
torque [tɔ:k]	n. 扭矩
rigidity [ri'dʒiditi]	n. 刚度
carbide ['kɑ:baid]	n. 碳化物

Technical Phrases

high-speed steel	高速钢
cemented carbide	硬质合金
CBN (cubic boron nitride)	立方氮化硼
machining center	加工中心
end mill	端铣刀
shell end mill	圆筒形端铣刀（或套筒铣刀）
contour milling	轮廓铣削
boring tool	钻削刀具，镗削刀具

Notes

1. ... As drill diameter and length increase, so does the error in hole size and location.

……随着钻头直径与长度的增加，孔径与孔位误差也随之增加。so 意为"也……"。

2. A reamer is a rotary end cutting tool used to accurately size and produce a good surface finish in a hole *which has been previously drilled or bored.*

过去分词短语 used to... 做 tool 的定语。which 引导的定语从句修饰 a hole。

Exercises

(1) Fill in the blanks according to the text with the words and phrases given below. Make changes if necessary.

  tap end mill spotting tool drill reamer boring tool

1) Shell end mills and _____ can perform a variety of machining operations.
2) The shortest _____ is always chosen so as to produce a hole of the required depth.
3) As an alternative to the center drill, _____ has a 90° included angle.
4) _____ is one of the most difficult machining operations to perform.
5) _____ are rotary end cutting tools used to accurately size and produce a good surface finish in a hole which has been previously drilled or bored.
6) Boring is usually performed with a single-point _____.

(2) Answer the following questions:

1) How many cutting tools are mentioned in the text? What are they?
2) What happens as drill diameter and length increase?
3) What are center drills used for?
4) Why is tapping one of the most difficult machining operations to perform?

【参考译文】

第 15 课 切 削 刀 具

在加工中心上精确加工零件，基本的要求是为每一步加工操作选择合适的刀具。但是，就每个具体工件的加工而言，选择刀具时一般没有完备的准则可供遵循。因此编程员必须熟练掌握切削刀具及其应用知识，以便对任何零件进行正确编程。

加工中心使用多种刀具完成各种加工操作。这些刀具有常规高速钢刀具、硬质合金嵌入式刀具、立方氮化硼复合刀片嵌入式刀具、多晶金刚石嵌入式刀具。常用刀具有端铣刀、钻头、丝锥、铰刀、镗刀等。

研究表明，在加工中心上，平均每个加工周期的工作时间中有 20% 进行铣削、10% 用于镗孔、70% 进行孔加工。普通铣床的切削时间大约占总加工时间的 20%，而加工中心的切削时间可高达 75%。结果是刀具利用率高了、寿命缩短了，导致刀具需求量大了。

端铣刀

端铣刀与空心端铣刀（见图 15-1）是加工中心上经常使用的两种刀具，它们能在加工中心上完成多种加工操作，例如铣削表面、铣槽与铣轮廓、刮孔口平面、反镗孔、用圆弧插补对孔进行粗加工与精加工。

钻头

通用钻头与专用钻头均可用于孔加工（见图15-2）。随着钻头直径与长度的增加，孔径误差与孔位误差也随之增加，因而孔加工时总是选用能加工出所需深度的最短钻头。若在加工中心上钻孔，则我们推荐使用粗短钻头。

中心钻

在孔加工中，可以先用中心钻（见图15-3）预加工出孔的准确位置，然后再进行其他加工。但是，使用中心钻进行孔加工有其缺陷：若使用不当，细小的导向钻头就容易崩断。这时可用定位工具代替中心钻，它有一个90°内角，常用于孔定位。

丝锥

在设计丝锥（见图15-4）时，应使丝锥能够承受攻丝所需转矩，并能清除孔内的加工屑。攻螺纹是最难进行的加工之一，主要原因如下：

- 排屑不彻底
- 切削液供给不足
- 不同材料，螺纹牙型的粗、细不同
- 攻螺纹速度及进给受螺纹导程影响
- 所需攻丝的深度

铰刀

铰刀（见图15-5）有多种型号与尺寸。铰刀是一种端部可旋转的切削刀具，主要对预先钻好或镗好的孔加工到准确尺寸，并得到良好的表面质量。

镗刀

镗孔是把预先钻过、镗过或用中心钻加工过的孔扩至准确尺寸与位置并达到所需表面粗糙度的一种操作。这项加工通常用镗刀完成（见图15-6）。在选取镗杆时，必须审慎考虑其长度与直径，因为随着长径比的增加，镗杆刚度会有所降低。比如，一个长径比为1:1的镗杆，其刚度比长径比为4:1的镗杆刚度高64倍。

Unit 16   Tooling Systems

Text

The machining center, a multifunction machine tool, uses a wide variety of cutting tools such as drills, taps, reamers, end mills, face mills, boring tools, etc., to perform various machining operations on a workpiece. For these cutting tools to be inserted into the machine spindle quickly and accurately, all these tools must have the same taper shank toolholders to suit the machine spindle. The most common taper used in NC machining center spindles is the No. 50 taper, which is a self-releasing taper. The toolholder must also have a flange or collar, for the tool-change arm to grab, and a stud, tapped hole, or some other device for holding the tool securely in the spindle by a power drawbar or other holding mechanism.

When one is preparing for a machining sequence, the tool assembly drawing is used to select all the cutting tools required to machine the part. Each cutting tool is then assembled off-line in a suitable toolholder and preset to the correct length. Once all the cutting tools are assembled and preset, they are loaded into specific pocket locations in the machine's tool-storage magazine where they are automatically selected as required by the part program. [1]

Tool Identification

NC machine tools use a variety of methods to identify the various cutting tools which are used for machining operations. The most common methods of identifying tools are:

1. *Tools pocket locations*

Tools for early machining centers were assigned a specific pocket location in the tool-storage magazine, and each tool was called up for use by the part program.

2. *Coded rings on toolholders*

A special interchange device reader was used to identify some tools by special coded rings on the toolholder.

3. *Tool assembly number*

Most modern MCUs have a tool identification feature which allows the part program to recall a tool from the tool-storage magazine pocket by using a five to eight-digit tool assembly number.

Each tool assembly number may be assigned a specific pocket in the tool-storage magazine by the tool data tape, by the operator using the MCU, or by a remote tool management console.

Tool Management Program

It is very important, to achieve the best productivity from any machine tool, to have a sound program which covers all aspects of cutting tools. [2] The best NC machine tool can not come anywhere close to its productivity potential unless the best cutting tools for each operation are available for use when they are required. The tool-management program must include such things as tool design, standard coding system, purchasing, good tooling practices, part programming which is cost effective, and the best use of cutting tools on the machine.

A good cutting tool policy must include the following:

1. *Standard policy*

1) A standard policy regarding cutting tools must be established. Every one should clearly understand the policy: the tool engineer, the part programmer, the supervisory staff, the setup person, and the machine tool operator.

2) The role that each person has in selecting the proper cutting tools must be clearly defined.

2. *Cutting tool dimensional standards*

1) All cutting tools purchased or specially made must conform to established cutting tool dimensional standards.

2) When it is necessary to recondition cutting tools, they should be ground to the next NC standard.

3) The part programmer must use the cutting tool standards for programming purposes.

3. *Rigid cutting tools*

1) Always select the shortest cutting tool possible for each job to ensure locational accuracy and rigidity.

2) Cutting tool holders should be of one-piece construction to provide rigidity.

4. *Tool preparation*

1) There must be a rigid policy on tool setting, compensation, and regrinding which is understood by everyone concerned.

2) Clearly define who has the responsibility for each so that there is no conflict or misunderstanding.

5. *Indexable insert tools*

1) Use cemented carbides insert-type tooling wherever possible because of their wear resistance, higher productivity, and dimensional accuracy.

2) Borazon (CBN) inserts should be used on hard ferrous metals where cemented carbides are not satisfactory.

3) Synthetic (polycrystalline) diamond inserts should be used for machining nonferrous materials.

The success or failure of a tool-management program depends in large part on the part programmer. To be most effective, the programmer must have a thorough knowledge of machining practices and procedures and the type of cutting tool required for each operation. Most modern CNC units have standard or optional features or programs available to make any tool-management policy more effective.

Technical Words

multifunction [ˌmʌltiˈfʌŋkʃən]　　　　n. 多功能
toolholder [ˈtuːlˌhəuldə (r)]　　　　n. 刀柄
flange [flændʒ]　　　　n. 边缘，法兰
collar [ˈkɔlə]　　　　n. 套环，卡圈，安装环

drawbar [ˈdrɔːbɑː]	n. 拉杆
console [kənˈsəul]	n. 控制台
regrind [riˈɡraind]	v. 重磨
ferrous [ˈferəs]	adj. 铁的
synthetic [sinˈθetic]	adj. 合成的
taper [ˈteipə]	n. 锥度
stud [stʌd]	n. 双头螺栓；销子；拉钉

Technical Phrases

hard ferrous metal	硬铁合金
nonferrous material	有色金属
coded ring	编码环
tool-storage	刀具存贮
dimensional standard	尺寸标准

Notes

(1) Once all the cutting tools are assembled and preset, they are loaded into specific pocket locations in the machine's tool-storage magazine where they are automatically selected as required by the part program.

一旦所有刀具装设就绪，便可装入机床刀库的指定刀位，加工时根据程序要求从刀库中自动选刀。

(2) It is very important, to achieve the best productivity from any machine tool, to have a sound program *which covers all aspects of cutting tools.*

句中 it 为形式主语，动词不定式 to have a sound program 是主句的真正主语；which 引出的定语从句修饰 program。

Exercises

(1) Place a "T" after sentences that are true and an "F" after those that are false.

1) All the tools used on NC machines must have the same taper shank toolholders.

2) All the cutting tools are assembled on-line in a suitable toolholders and preset to the correct length.

3) Each tool used on the machine has an assembly number and a specific pocket in the tool-storage magazine.

4) Because the cutting tools are used during the machine operation, only the machine tool operator should know the standard policy regarding cutting tools.

5) The success or failure of a tool-management program depends in large part on the part programmer.

(2) Fill in the blanks according to the text with the phrases given below. Make changes if necessary.

tool-storage magazine   tool identification   assembly number
tool assembly drawing   shank toolholder

A wide variety of cutting tools are used on the machining center. Each machine tool has the same tape _____ to suit the machine spindle. The _____ is used to select all the required cutting tools when a machining sequence is prepared. To identify each tool, we can use three methods. Most modern MCUs have a (an) _____ feature which allows the part program to recall a tool from the _____. Each tool _____ is assigned a specific pocket in the tool-storage magazine.

(3) Answer the following questions:

1) How can all the cutting tools be inserted into the machine spindle quickly and accurately?

2) How many methods does an NC machine tool use to identify the various cutting tools, and what are they?

3) Which kind of cutting tool should we select to ensure locational accuracy and rigidity?

4) Why do modern CNC units have to make any tool-management policy more effective?

【参考译文】

第16课 刀具系统

加工中心是一种多功能机床，可用钻头、丝锥、铰刀、端铣刀、平面铣刀、镗刀等各种刀具对工件进行多种加工。为了使这些刀具能迅速准确地插入机床主轴，所有刀具必须有相同的锥柄与机床主轴适配。数控加工中心上最通用的主轴锥柄是 No.50，这种锥柄具有自释放功能。刀柄还必须有凸沿和卡圈，以便换刀机械手夹持；还应有专用拉钉、螺孔或其他装置，通过电动拉杆或其他夹紧机构将刀具牢固夹紧在主轴上。

准备加工工序时，要用刀具装配图选择加工零件所需的全部刀具，然后将每把刀具预先装配到合适的刀柄上，并预设正确长度。一旦所有刀具装设就绪，便可装入机床刀库的指定刀位，加工时根据程序要求从刀库中自动选刀。

刀具识别

数控机床使用多种方法识别加工用的各种刀具，最常用的刀具识别方法有：

1. 刀具刀位法

早期加工中心在刀库中给每把刀具分配一个指定的刀位，零件程序可以调用任何刀具。

2. 刀柄编码环法

由交互式识读机根据刀柄上的编码环来识别刀具。

3. 刀具装配号法

多数现代机床控制单元允许零件程序使用 5~8 位刀具号码从刀库中调用刀具。

每把刀具的号码都与刀库中的一个特定存储位置对应，其位置可由穿孔带设定，也可由操作人员用机床控制单元设定，或通过刀具管理遥控平台设定。

刀具管理程序

为了达到最大生产能力，给机床配备功能齐全的刀具管理软件就显得十分重要。只有每步加工都能用上最合适的刀具，性能最优越的数控机床才有可能接近其潜在加工能力。刀具管理软件应涵盖多种信息：刀具类型、标准代码体系、外购状况、行之有效的加工技巧、有效的零件编程、刀具在机床上的最佳应用等。

良好的选刀原则应包括以下内容：

1．标准规则

1）必须建立刀具的标准规则。从刀具工程师、工件编程员、高级管理人员到安装工、操作工，人人都应该弄清楚这个规则。

2）必须明确每个人在选刀时担当的角色。

2．刀具尺寸标准

1）所有外购、特制刀具必须符合公认的刀具尺寸标准。

2）必须重磨刀具时，应将其修磨到下一级 NC 尺寸标准。

3）编程员必须按刀具标准准则进行编程。

3．刚性切削刀具

1）尽可能选用最短刀具完成每项工作，以确保定位精度与刚度。

2）刀柄应为一体化结构，以提高刚度。

4．刀具准备

1）在刀具安装、刀具补偿、重磨刀具时，相关人员必须严格遵守有关规定。

2）必须明确每项工作中的个人分工，确保没有争议或误解。

5．可转位刀片刀具

1）尽可能使用硬质合金刀具，这种刀具抗磨损性好、生产效率高、尺寸精度高。

2）立方氮化硼刀片用于硬质合金刀具无法满足加工要求的硬铁材料加工。

3）多晶刀片用于加工非铁材料。

刀具管理程序的好坏在很大程度上取决于工件编程员。为使工作更有效，编程员必须全面掌握实际加工知识、工序及每种加工所选刀具类型。为使刀具管理规则能更有效地发挥作用，大多数现代 CNC 装置都带有标准功能、选配功能及有关程序。

【Reading Material】

Set Up Procedures

The following sequence of operations is strongly recommended for setting up the machine:

(1) Load a program into memory. This is either manually entered or downloaded from a CAM package via the RS-232 interface.

(2) Determine the tools needed and get them ready.

CAUTION! Do not exceed the Maximum Specifications!

CAUTION!

- Extremely heavy tool weights should be distributed evenly.
- Ensure there is adequate clearance between tools in the tool changer before running an automatic operation.

(3) Use a vise or fixture to hold the part and mount it on the mill.

(4) Locate the X and Y zero points of your program on the part. Usually these points will coin-

cide with the print reference point where dimensioning begins and needs to be clearly indicated by the programmer. Use an edge finder or indicator to locate this point with the handle function. After locating the programmed zero point, push the DISPLAY OFFSET key and PAGE DOWN until the work zero offset page appears. Move the cursor down to G54 X, the work zero offset. Push the PART ZERO SET button and the X-axis machine value will be stored at this location. Move the cursor to the G54 Y location and repeat the above. You have now told the machine where part zero is located. Usually Z and A values will not have to be set and should be zero.

(5) Remove any tools from changer and MDI a T1 M6 command to install tool #1 into spindle (it should be empty). Put your tool #1 into the spindle using the TOOL RELEASE button. Push the OFSET key and PAGE DOWN to get to the tool offset page and cursor to tool #1. Do not install any tools directly into the carousel. Use MDI or ATC FWD/REV to retrieve tools.

(6) Push the Z-JOG key until you are close to the top of your part. (The top of your part should be Z0). Use the handle to accurately position the tool edge to Z0. Push the TOOL OFSET MESUR key and the Z machine value will be stored in tool-offset #1. Note that this automatic offset measurement works with G43 only and the work Z offset must be zero.

(7) Push the NEXT TOOL key and the Z-axis will retract to tool change position and tool #2 (empty) will be installed in the spindle. Put your tool #2 into the spindle and jog to Z zero as you did for tool #1. The cursor will automatically be on offset #2. Push TOOL OFSET MESUR.

(8) Repeat this procedure until all tools are measured and installed.

(9) MDI a T1 M6 to return to tool #1.

(10) You are now ready to run your program.

(11) Please note that in order to load and measure all of your tools, you do not need to use any keys other than JOG, TOOL OFSET MESUR, and NEXT TOOL. Also note that this automatic offset measurement works with G43 only.

Unit 17　Adaptive Control

Text

　　The most important factors affecting the efficiency of machining on a machining center are cutter speed, feed, and depth of cut. If a cutting tool is run too slowly, valuable time will be wasted, resulting in lost production. Too high a speed will create too much heat and friction at the cutting edge of the tool, which quickly dulls the cutter. This results in having to stop the machine to either recondition or replace the cutter. Somewhere between these two extremes is the efficient cutting speed for each material being cut.

　　Feed is the rate at which the work is fed into a revolving cutter. If the work is fed too slowly, time will be wasted, resulting in lost production, and cutter chatter, which shortens the life of the cutter, may occur. [1] If work is fed too fast, the cutter teeth can be broken. Much time will be wasted if several shallow cuts are taken instead of one deep cut or roughing cut. Therefore, speed, feed, and depth of cut are three important factors which affect the life of the cutting tool and the productivity of the machine tool. Charts and tables containing speeds, feeds, and depth of cut are available from cutting tool manufacturers, machinery handbooks, etc.

　　The programmer must select the proper speeds and feeds for each part to be machined, so that the part is produced in the shortest period of time while considering cutting tool life. Generally, cutting tool breakage occurs because the cutter is dull, or the depth of cut is changed because of variances in workpiece thickness. [2] Whenever a cutting tool becomes dull or is broken, the NC machine must be stopped to recondition or replace the cutting tool. To get the best productivity from an NC machine tool, optimum speeds and feeds should be used for machining operations.

　　A feature that is fast becoming popular is that of torque control machining, the torque being calculated from measurements at the spindle drive motor. This device will increase productivity by preventing or sensing damage to the cutting tool. The torque is measured when the machine is turning but not cutting, and this value is stored in the computer memory.

　　As the machining operation begins, the stored value is subtracted from the torque reading at the motor. This will give the net cutting torque, which is compared to the programmed torque or limits stored in the computer (or on NC tape). If the net cutting torque exceeds the programmed torque limits, the computer will act by reducing the feed rate, turning on the coolant, or even stopping the cycle. The feed rate will be lowered whenever the horsepower requirements exceed the rated motor capacity or the programmed code value.

　　The system display of three yellow lights advises the operator of the operational conditions in the machine at the time. A left-hand yellow light indicates that the torque control unit is in operation. The middle yellow light indicates that the horsepower limits are being exceeded. The right-hand light comes on when the feed rate drops below 60 percent of the programmed rate. The meter indicates the cutting torque (or operational feed rate) as a percent of the programmed feed rate.

As the tool gets dull, the torque will increase and the machine will back off on the feed and ascertain the problem. Excessive material could be on the workpiece, or a tool might be very dull or broken. If the tool is dull, the machine will finish the operation and a new backup tool of the same size will be selected from the storage chain when that operation is performed again. If the torque is too great, the machine will stop the operation on the workpiece and program the next piece into position for machining

Technical Words

adaptive [əˈdæptiv]   adj. 适应的
recondition [ˈriːkənˈdiʃən]   v. 使复原
chatter [ˈtʃætə]   n. 刀具振颤
handbook [ˈhændˌbuk]   n. 手册
coolant [ˈkuːlənt]   n. 切削液
ascertain [ˌæsəˈtein]   v. 确定；查明

Technical Phrases

adaptive control   适应控制
revolving cutter   旋转刀具
torque control machining   转矩控制加工
spindle drive motor   主轴驱动电动机

Notes

(1) If the work is fed too slowly, **time will be wasted**, resulting in lost production, **and cutter chatter**, *which shortens the life of the cutter*, **may occur.**

此句主干为：time will be wasted, and cutter chatter may occur。which 引导的定语从句修饰 cutter chatter（刀具振颤）。

(2) Generally, cutting tool breakage occurs because **the cutter is dull**, or **the depth of cut is changed** because of variances in workpiece thickness.

引起 cutting tool breakage 的原因有二，故 because 后有两个从句：the cutter is dull 和 the depth of cut is changed。第二个原因是"工件厚度变化"所致，用介词短语 because of ... 加以说明。注意：because 后接从句，而 because of 后接短语。

Exercises

(1) Place a "T" after sentences that are true and an "F" after those that are false.

1) Since too slow a speed will waste valuable time, a cutting tool should run as fast as possible.

2) If a work is fed too slowly, cutter chatter would occur, which shortens the life of the cutter.

3) The efficient cutting speeds, feeds, and depth of cut can only be found in machinery handbook.

4) Cutting tool breakage occurs only when the cutter is dull.

(2) Fill in the blanks according to the text with the words given below. Make changes if necessary.

    torque   limit   feed   programmer   productivity

Cutter speed, _____, and depth of cut are the three important factors that affect the efficiency of machining on a machining center. It is the _____ who must select the proper speeds and feeds for each part to be machined. To get the best _____ from an NC machine tool, optimum speeds and feeds should be used. Nowadays, _____ control machining is becoming popular. If the net cutting torque exceeds the programmed torque _____, the computer will act to lower it.

(3) Answer the following questions:
1) If the work is fed too fast, what will happen?
2) How can we get the best productivity from NC machine tools?
3) How can we calculate the net cutting torque?
4) If there is excessive material on the workpiece, what will happen to the cutting tool?

【参考译文】

第 17 课　自适应控制

影响加工中心加工效率的最主要的因素有切削速度、进给量及背吃刀量。如果进刀太慢，就会浪费宝贵时间，导致产量下降；若切削速度太高，摩擦产生过多热量会使刀刃很快变钝，结果只能停车，重修或更换刀具。因此，这两种极端情况间必有适合每一种材料的有效切削速度。

进给量是指刀具旋转一周时工件的移动量。若工件进给太慢，则会浪费时间，导致产量下降，并可能发生刀具振颤，缩短刀具寿命。如果工件进给太快，则可能崩坏刀齿。若采用多次浅切而不是一次深切或粗切，就会浪费大量时间。因此，切削速度、进给量及背吃刀量便构成影响刀具寿命与机床生产效率的三大要素。至于切削速度、进给量及背吃刀量的具体参数，可从刀具制造商家得到，亦可查询机械手册等资料。

编程人员必须为每个待加工工件选择合适的切削速度与进给量，以便在考虑刀具寿命的情况下，用最短的时间完成工件加工。发生崩刀通常是因刀具变钝或工件厚度变化引起背吃刀量改变造成的。每当刀具变钝或崩断时，数控机床就会停止运行，以便重新修磨或更换刀具。要使数控机床获得最佳加工效率，就应选用最优切削速度与进给量。

转矩控制加工是目前迅速推广应用的一种自适应控制方法，而转矩则是根据主轴电动机测量数据计算得出的。这种方法可以检测并防止刀具损伤，从而可提高生产效率。该转矩是在机床运转但不切削的状态下测量的，其数值被存于计算机存储器中。

加工之初，先从电动机转矩值中减去已存储的转矩值，从而得到净切削转矩；再把净切削转矩与程序设定转矩或计算机（或穿孔带）中存储的转矩极限作对比，若净切削转矩超出程序设定的转矩极限，计算机就起作用：降低进给量、打开切削液，甚至终止加工循环。

每当所需功率超过电动机额定容量或程序设定值时，进给率就会有所降低。

该系统设有三个指示黄灯，向操作人员提醒加工状况。左边的黄灯点亮，则表明转矩控制装置处于工作状态。中间的黄灯点亮，则表明功率超出了限度。当进给率降到程序设定值的60%以下时，右边的黄灯点亮。仪表显示的切削转矩（或加工进给量）是以程序设定进给量的百分率形式给出的。

刀具变钝时，转矩就会增大，机床就会停止进给，并查找故障。故障可能是工件上切削量过大、刀具变钝或崩断所致。如果是刀具变钝，机床就会终止加工，并从刀库中选取同样尺寸的备用刀具，再继续加工。如果转矩太大，机床则会停止加工该工件，并开始加工下一个工件。

Unit 18　Wire-Cut EDM

Text

Principle of EDM

Electrical discharge machining, commonly known as EDM, is a controlled metal removal process whereby an electric spark is used to cut (erode) the workpiece, which then takes the shape opposite to that of the cutting tool or electrode.[1] The electrode and the workpiece are both submerged in a dielectric fluid, which is generally a light lubricating oil. This dielectric fluid should be a nonconductor of electricity. A servomechanism maintains a gap of about 0.0005 to 0.001 in. (0.01 to 0.02mm) between the electrode and the workpiece, preventing them from coming into contact with each other. A direct current of low voltage and high amperage is delivered to the electrode at the rate of approximately 20,000 hertz (Hz). These electrical energy impulses become sparks which jump the gap between the electrode and the workpiece through the dielectric fluid.[2] Intense heat is created in the localized area of the spark impact; the metal melts and a small particle of molten metal is expelled from the workpiece. The dielectric fluid, which is constantly being circulated, carries away the eroded particles of metal and also helps in dissipating the heat caused by the spark.

There are two types of EDM machines used in industry: the vertical EDM machine (Fig18-1a) and the wire-cut EDM machine (Fig. 18-1b). Since the wire-cut EDM is generally used for machining complex forms which require NC programming, only this type will be discussed in detail.

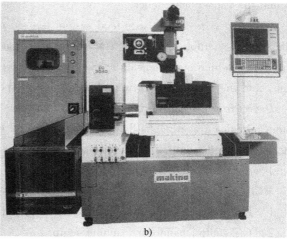

Fig. 18-1　Electrical discharge machines
a) Vertical EDM　b) Wire-cut EMD

The Introduction of Wire-Cut EDM

The wire-cut EDM is a discharge machine which uses NC movement to produce the desired contour or shape on a part. It does not require a special-shaped electrode; instead, it uses a continuous traveling wire under tension as the electrode. The electrode or cutting wire can be made of brass, copper, or any other electrically conductive material ranging in diameter from 0.002 to 0.012 in. (0.05 to 0.30mm). The path that the wire follows is controlled along a two-axis (XY) contour, eroding (cutting) a narrow slot through the workpiece.[3] This controlled movement is continuous and simultaneous in increments of 0.00005 in. (0.0012mm). Any contour may be cut to a high degree of accuracy and is repeatable for any number of successive parts. A dielectric fluid, usually deionized water which is constantly being circulated, carries away the eroded particles of metal.[4] The dielectric fluid maintains the proper conductivity between the wire and the workpiece and assists in reducing the heat caused by the spark.

Parts of the Wire-Cut EDM

The main parts of the wire-cut EDM are the bed, saddle, table, column, arm, UV axis headstock, wire feed and dielectric systems, and machine control unit (MCU) (Fig. 18-2).

Bed: The bed is a heavy, rugged casting used to support the working parts of the wire-cut EDM. Guide rails ("ways") are machined on the top section, and these guide and align major parts of the machine.

Saddle: The saddle is fitted on top of the guide rail and may be moved in the XY direction by feed servomotors and ball lead screws.

Table: The table, mounted on top of the saddle, is U-shaped and contains a series of drilled and tapped holes equally spaced around the top surface. These are used for the workpiece holding and clamping devices.

Column: The column, which is attached to the bed, supports the wire feed system, the UV axes, and the capacitor switches.

Wire feed system: The wire feed system is used to provide a continual feed of new wire (electrode) for the cutting operation. The wire is fed from a supply spool through a series of guides and guide rollers which apply tension to the wire[5]. The wire travels in a continuous path past the workpiece, and the used wire is rewound on a take-up spool.

Dielectric system: The dielectric system contains filters, an ion exchanger, and a cooler. This system provides a continuous flow of clean deionized water at a constant temperature. The deionized water stabilizes the cutting operation, flushes away particles of electrode and workpiece material that have been eroded, and cools the workpiece.

MCU: The MCU can be separated as three individual panels.

1) The control panel for setting the cutting conditions.

2) The control panel for machine setup.

3) The control panel for manual data input (MDI) and cathode-ray tube (CRT) character display (Fig. 18-3).

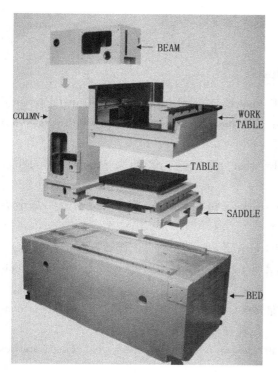

Fig. 18-2 The main parts of a wire-cut EDM

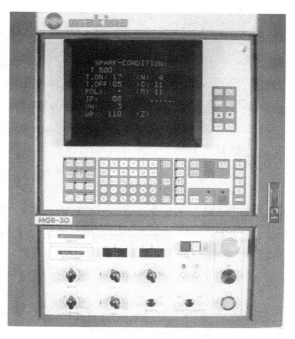

Fig. 18-3 Control panel

Technical Words

spark [spɑːk]	n.	火花
erode [iˈrəud]	v.	腐蚀，电蚀
submerge [səbˈməːdʒ]	v.	淹没，浸入
nonconductor [ˌnɔnkənˈdʌktə]	n.	绝缘体
amperage [ˈæmpɛəridʒ]	n.	安培数
discharge [disˈtʃɑːdʒ]	v.	放电
dissipate [ˈdisipeit]	v.	驱散；耗散
dielectric [ˌdaiiˈlektrik]	n.	电介质；绝缘体
deionize [diːˈaiənaiz]	v.	除去离子
rugged [rʌgid]	adj.	结实的，坚固的
align [əˈlain]	v.	调整，校准
clamp [ˈklæmp]	v.	夹住
ion [ˈaiən]	n.	离子
capacitor [kəˈpæsitə]	n.	电容器
expel [iksˈpel]	v.	驱逐，开除，排出，发射

Technical Phrases

CRT   阴极射线管

Wire-Cut EDM　　　　　　　　　　　　电火花线切割加工
guide roller　　　　　　　　　　　　导轮

Notes

(1) Electrical discharge machining, commonly known as EDM, is a controlled metal removal process whereby an electric spark is used to cut (erode) the workpiece, *which then takes the shape opposite to that of the cutting tool or electrode.*

句中关系副词 whereby 相当于 by which 或 by what，意思是"借以"。which 引出的非限定性定语从句修饰 the workpiece。

(2) These electrical energy impulses become sparks *which jump the gap between the electrode and the workkpiece through the dielectric fluid.*

句中 which 引导的定语从句修饰 sparks。dielectric fluid 可译为"工作液"。

(3) The path *that the wire follows* is controlled along a two-axis (XY) contour, eroding (cutting) a narrow slot through the workpiece.

句中 that the wire follows 为定语从句，修饰 The path。句子的主干是：The path is controlled along a two-axis (XY) contour. eroding... 是现在分词短语作伴随状语。

(4) A dielectric fluid, usually deionized water *which is constantly being circulated*, carries away the eroded particles of metal.

deionized water 意为"除去离子的水"。which is constantly being circulated 为定语从句，修饰 deionized water。句子主干是：A dielectric fluid carries away the eroded particles of metal.

(5) The wire is fed from a supply spool through a series of guides and guide rollers *which apply tension to the wire.*

which 引导的定语从句修饰 guides and guide rollers，说明它们所起的作用。supply spool 意为"贮丝筒"；apply... to... 意为"把……施加于……"。

Exercises

(1) Place a "T" after sentences that are true and an "F" after those that are false.

1) EDM uses an electric spark to cut the workpiece into the same shape as the electrode has.

2) The dielectric fluid is used only to carry away the eroded particles of metal and assist in reducing the heat caused by the spark.

3) A wire-cut EDM machine can't cut a contour to a high degree of accuracy.

4) The dielectric system of a wire-cut EDM provides a continuous flow of clean deionized water at a constant temperature in order to keep the operation done successfully.

(2) Fill in the blanks according to the text with the words given below. Make changes if necessary.

　　　　particle　　melt　　spark　　shape　　dielectric　　electrode

Unlike other machines, EDM uses an electrical _____ to cut the workpiece. The electrode and the workpiece are both submerged in a _____ fluid. A servomechanism keeps a gap between the _____ and the workpiece. The spark impact creates intense heat to _____ the metal. The dielectric fluid carries away the eroded _____ of metal and assists in reducing the heat caused by the spark. As the wire travels in a continuous path through the workpiece, it will take the _____ opposite to that of the electrode.

(3) Answer the following questions:
1) How can EDM remove metal from a workpiece?
2) Why must an appropriate gap be maintained between the electrode and the workpiece?
3) What is the dielectric fluid used for?
4) What kind of metal can be used as an electrode?
5) How many main parts does a wire-cut EDM have?

【参考译文】

第18课 电火花线切割加工

电火花加工原理

电火花加工通常简称 EDM，它是采用电火花放电腐蚀工件来控制金属切削的加工方法。这样加工出的工件形状与切割刀具即电极的形状相反。加工时，电极与工件都浸入工作液中，常用的工作液是轻润滑油，此液体应是电的绝缘体。伺服机构保持电极与工件之间有大约 0.0005~0.001 英寸（0.012~0.025mm）的间隙，以免二者互相接触。给电极加以频率约为 20 000Hz 的低压大电流脉冲序列，脉冲能量在电极与工件间的工作液中火花放电，从而在放电局部产生大量热量而使金属熔化，其微粒从工件上脱落下来。持续循环的工作液不仅冲走熔化后的金属微粒，并且有助于散发电火花产生的热量。

工业中使用的电火花加工机床分为两类：立式主轴电火花加工机床（见图 18-1a）与电火花线切割加工机床（见图 18-1b）。由于电火花线切割加工通常用数控编程来加工复杂形状，此处只详细介绍这类机床。

电火花线切割加工介绍

电火花线切割加工是一种放电加工方式，加工时用数控运动在零件上产生所需轮廓或形状。它无需特制形状的电极，而是用连续运动且处于张紧状态的金属线作为电极。电极即切割线，它可以用黄铜线、铜线或其他导电材料制成，其直径在 0.002~0.012 英寸（0.05~0.30mm）之间。切割线被控制在 X-Y 平面内沿工件轮廓曲线移动，从而在工件上电蚀出一条狭缝。切割线的这种运动是以 0.00005 英寸（0.0012mm）的微小增量连续进行的。这种方法可以非常准确地加工出任何轮廓形状，并可连续性重复加工许多零件。工作液通常是去除离子的水，它不断循环流动，冲走腐蚀的金属微粒。工作液不仅可以使切割线与工件之间保持合适的导电性，并且有助于耗散火花产生的热量。

电火花线切割加工机床的组成部件

电火花线切割加工机床的主要部件有：床身、床鞍、工作台、立柱、臂、UV 床头箱、切割线进给与电解系统、机床控制单元（MCU）（见图 18-2）。

床身：床身是较重的刚性铸件，用于支撑线切割机床的工作部件。床身上面装有导轨，对机床的主要部件起导向与调整作用。

床鞍：床鞍装在导轨上面，在进给伺服电动机与滚珠丝杠的作用下可沿 XY 方向运动。

工作台：U 形工作台装在床鞍上面，上表面均匀分布着一系列孔或螺纹孔，用于装夹工件。

立柱：立柱与床身相连，为线进给系统、UV 轴及电容开关提供支撑。

切割线进给系统：切割线进给系统为线切割加工不断提供新线（电极）。新线从贮丝筒的供线端绕过一系列导轮与导柱，这些导轮和导柱给切割线提供一定张紧力。切割线沿连续路径穿过工件而运动，用过的线又绕在贮丝筒的收线端。

工作液系统：工作液系统包括过滤器、离子交换器及冷却器。该系统可提供连续流动的恒温去离子水。去离子水能起到稳定切削状态、冲洗电极与工件上电蚀微粒以及冷却工件的作用。

机床控制单元：机床控制单元可分为三个独立控制面板：

1）加工条件控制面板。

2）机床部件控制面板。

3）手动数据输入（MDI）及阴极射线管（CRT）字符显示控制面板（见图 18-3）。

Unit 19　CNC Lathes

Text

A CNC lathe(See Fig. 19-1) is a machine tool used primarily for producing surfaces of revolution. Its main function is to create high-quality cylindrical parts in a minimum amount of time. Besides, it can also machine other operations, such as drilling, tapping, boring, and threading.

Fig. 19-1　CNC lathe

Unlike the conventional types of machining tools, which generally take up a lot of space and require a huge amount of time to produce outputs, the CNC lathe is relatively smaller and does not need highly-skilled and trained machinists.[1] They're easy to operate, and anyone can pretty much navigate the different tools and controls as soon as the tool is set up.

The working principle of the CNC lathe is the same as the engine lathe. In turning the workpiece a cutter moves in the direction parallel to the axis of rotation of the workpiece or at an angle to this axis, cutting off the metal from the surface of the workpiece. This movement of the cutter is called the feed. The cutter is clamped in the tool post which is mounted on the carriage.[2] The carriage is the mechanism feeding the cutter in the needed direction.

Bed

The lathe bed is the main frame, supporting components of the machine. It is usually made of grey or nodular cast iron to damp vibrations. It has guide-ways to allow the carriage to slide easily lengthwise. The bed lies either flat or at a slant to accommodate chip removal.[3] The slant of the bed is usually 30 to 45 degrees. The height of the lathe bed should be appropriate to enable the technician to do his or her job easily and comfortably.

Headstock

The headstock is fixed at the left side of the lathe bed and includes the spindle whose axis is parallel to the guide-ways. The spindle is always hollow, i. e. , it has a through hole extending lengthwise. Bar stocks can be fed through that hole if continuous production is adopted. Also, that hole has a tapered surface to allow mounting a plain lathe center. The outer surface of the spindle is threaded to allow mounting of a chuck, a face plate, or the like. The spindle which rotates the workpiece is driven by a motor. This motor, which is programmable in revolutions per minute,[4] delivers the required horsepower and torque through a driving belt or series of gears.

Tailstock

The tailstock assembly consists basically of three parts, its lower base, an intermediate part, and the quill. The lower base is a casting that can slide on the lathe bed along the guide-ways,[5] and it has a clamping device to enable locking the entire tailstock at any desired location, depending upon the length of the workpiece. The intermediate part is a casting that can be moved transversely to enable alignment of the axis of the tailstock with that of the headstock. The third part, the quill, is a hardened steel tube, which can be moved longitudinally in and out of the intermediate part as required. The hole in the open side of the quill is tapered to enable mounting of lathe centers or other tools like twist drills or boring bars. The quill can be locked at any point along its travel path by means of a clamping device.

The tailstock is used to support one end of the workpiece. Two types of tailstocks are available on CNC turning machines: manual and programmable. The manual tailstock is moved into position by the use of a switch or hand wheel. The programmable tailstock can be moved manually or can be programmed like the tool turret.

Tool Turrets

Tool turrets on turning machines come in all styles and sizes,[6] but the basic function of the turret is to hold and quickly index cutting tools. Each tool or tool position is numbered for identification. During tool changing, the turret moves to a clearance position, bringing the new tool into the cutting position. Most tool turrets can move bi-directionally to assure the fastest tool indexing time. Tool turrets can also be indexed manually, using a button or switch located on the control panel.

Fixtures

Fixtures are work-holding devices used for odd-shaped or other hard to hold workpieces. Fixtures may be held in the chuck or can be bolted directly to the spindle. Fixtures are spun in the spindle, so they must be balanced. Unbalanced fixtures can cause severe damage to the machine and possible injury to the operator.

Technical Words

tapping ['tæpiŋ]	n.	攻螺纹
boring ['bɔːriŋ]	n.	镗削，镗孔
thread [θred]	n.	螺纹
	v.	攻螺纹
clamp [klæmp]	n.	夹持；夹钳
	v.	夹持，夹住
carriage ['kæridʒ]	n.	支架，底座；拖架
vibration [vai'breiʃən]	n.	振动
headstock ['hedstɔk]	n.	床头箱，主轴箱
chuck [tʃʌk]	n.	卡盘夹住；夹头；夹具，夹盘
torque [tɔːk]	n.	转矩；力矩；扭矩
gear [giə]	n.	齿轮；传动装置
tailstock ['teilstɔk]	n.	尾座；尾架
quill [kwil]	n.	套筒轴
taper ['teipə]	n.	锥形物

	v.	使成锥形
fixture ['fikstʃə]	n.	夹具，固定装置
bolt [bəult]	n.	螺栓
	v.	用螺栓固定
spin [spin]	n.	旋转
	v.	使旋转

Technical Phrases

clamping device	夹持装置，定位装置
tool post	刀座
twist drill	麻花钻
boring bar	镗杆；钻杆
turning machine	车削中心；车床
manual tailstock	手动尾架；普通尾架
programmable tailstock	程控尾架；自动尾架
tool turret	转塔刀架
cutting tool	切削工具，刀具
tool changing	换刀
control panel	控制面板
work-holding device	夹具

Notes

(1) Unlike the conventional types of machining tools, *which generally **take up** a lot of space and **require** a huge amount of time to produce outputs*, the CNC lathe is relatively smaller and does not really need highly-skilled and trained machinists.

句中 which generally take up a lot of space and require a huge amount of time to produce outputs 为非限定性定语从句，修饰 the conventional types of machining tools（传统机床）；从句谓语动词有两个，一个是 take up，一个是 require；a lot of space 意为"许多空间"，a huge amount of time 意思是"大量时间"；highly-skilled 可译为"技术娴熟的"。

(2) The cutter is clamped in the tool post *which is mounted on the carriage.*

本句中，which is mounted on the carriage 为定语从句修饰 the tool post（刀座）。

(3) The bed lies either flat or at a slant to accommodate chip removal.

句中的 to accommodate chip removal 可译为"以便清除碎屑（排屑）"；at a slant 可译为"倾斜地"。

(4) This motor, *which is programmable in revolutions per minute*, ...

句中，which is programmable in revolutions per minute 为非限定性定语从句，修饰主语 This motor；这里的 revolution 一词意思是"转"，revolutions per minute 意思是"每分钟的转数"。

(5) The lower base is a casting *that can slide on the lathe bed along the guide-ways* ...

句中，that can slide on the lathe bed along the guide-ways 为定语从句，修饰 a casting（铸件）。

(6) Tool turrets on turning machines come in all styles and sizes...
此处的 come 为"使用，供应"之意。

Exercises

(1) Place a "T" after sentences that are true and an "F" after those that are false.

1) A CNC lathe is a machine tool used primarily for producing surfaces of revolution and flat edges.
2) The CNC lathe can not machine drilling and boring operations.
3) The CNC lathe really need highly-skilled and trained machinists.
4) The lathe bed lies either flat or at a slant to accommodate chip removal.
5) The basic function of the turret is to hold and quickly index cutting tools.

(2) Fill in the blanks according to the text with the words given below. Make changes if necessary.

hollow   turret   headstock   fixture   tailstock   nodular

The CNC lathe bed is usually made of grey or _____ cast iron to damp vibrations. The _____ is fixed at the left hand side of the lathe bed. The spindle is always _____. The _____ is used to support one end of the workpiece. The basic function of the _____ is to hold and quickly index cutting tools. _____ are work-holding devices used for odd-shaped or other hard to hold workpieces. In some sense, the structure of the CNC lathe is the same as the engine lathe.

(3) Answer the following questions:

1) What is the main function of a CNC lathe?
2) Can a CNC lathe machine drilling, tapping, boring, and threading operations?
3) What is the function of the lathe bed?
4) Why is the spindle always hollow?
5) How many parts does the tailstock assembly consist of basically? What are they?
6) What is the basic function of the turret?
7) What are the fixtures used for?

【参考译文】

第19课　数控车床

数控车床（见图19-1）主要是用于加工旋转表面的机床。数控车床的主要功能是以最短时间加工出优质圆柱形零件，另外，它也可以进行钻削、攻螺纹、镗削、车螺纹等加工操作。

传统机床占用空间大、加工零件时间长。与传统机床比，数控车床相对比较小，也无需技能高超、专门训练的技师来操作。数控车床操作便捷，只要装夹好刀具，任何人都能轻松控制不同的刀具并操纵机床。

数控车床与普通车床工作原理是一样的。车削工件时，刀具要么沿平行于工件旋转轴的方向运动，或与加工轴成一定角度进行运动，以切除工件表面的金属。刀具的这种运动叫做进给。刀具安装在刀座上，刀座安装在刀架上，刀架是沿着需要的方向进给的机构。

床身

床身是机床的基体，用来支撑机床的其他部件。床身通常用灰铸铁或球墨铸铁制成，以便减振。床身装有导轨，以便于刀架纵向滑动。床身布局分为平床身和斜床身，斜床身排屑方便。通常斜床身倾斜30°~45°。床身高度以机械师操作方便舒适为宜。

主轴箱

主轴箱固定在床身左侧，箱内装有主轴，主轴与导轨平行。主轴往往是空心的，即其中有一个贯穿全长的通孔，若连续生产，可以通过该孔输送棒料。同时，此孔为锥形表面，可安装普通车床顶尖。主轴外表面有螺纹，可安装卡盘、花盘或类似装置。电动机驱动主轴使工件旋转；电动机转速用每分钟多少转度量；电动机通过传动带或齿轮组提供动力和转矩。

尾架

尾架基本上由三部分组成：底座、中间部分及套筒轴。底座属铸件，可在床身沿导轨滑动；底座配有夹紧装置，能使整个尾架根据工件长度锁定于任何位置。中间部分也是铸件，可横向移动，确保尾架轴线与主轴箱轴线共线。第三部分是套筒轴，是经淬硬处理后的钢管，需要时可在中间部分纵向进出。套筒轴开口端有一锥形孔，可安装车床顶尖或诸如麻花钻和镗杆之类的其他工具。套筒轴通过夹紧装置可沿其移动路径被锁定在任何位置上。

尾架用于支撑工件的一端。有两种尾架可供数控车床使用，一个是手动尾架，另一个是自动尾架。手动尾架使用开关或手轮将尾架送入工位；自动尾架既可以人工操作，也可以像转塔刀架那样程控操作。

转塔刀架

车床用的转塔刀架，型号、规格一应俱全，但其基本功能不外乎夹持刀具和快速寻刀。每把刀、每个刀位都已编号，以便识别。换刀过程中，转塔刀架先移至某一空位，再把新刀输送到切削位置。大多数转塔刀架可双向运动，确保以最短时间检索刀具。也可运用控制面板上的按钮或开关手动操作转塔刀架。

夹具

夹具用来夹持形状不规则或难以夹持的工件。夹具可以固定在卡盘上，或直接用螺栓固定在主轴上。夹具与主轴一起转动，因此二者必须平衡。不平衡的夹具会对机床造成严重损害，甚至可能伤及操作人员。

【Reading Material】

CNC Lathe Cutting Tools

The shape and geometry of the lathe tools depend upon the purpose for which they are employed. Lathe cutting tools include the following types:

Turning Tools

Turning tools can be either finishing or rough turning. Rough turning tools have small nose radii and are employed when deep cuts are made. On the other hand, finishing tools have larger nose radii and are used for obtaining the final required dimensions with good surface finish by making slight depths of cut. Rough turning tools can be right-hand or left-hand types, depending upon the direction of feed. They can have straight, bent, or offset shanks.

Facing Tools

Facing tools are employed in facing operations for machining plane side or end surfaces. There are tools for machining left-hand-side surfaces and tools for right-hand-side surfaces. Those side surfaces are generated through the use of the cross feed, contrary to turning operations, where the usual longitudinal feed is used.

Cutoff Tools

Cutoff tools, which are sometimes called parting tools, serve to separate the workpiece into parts and/or machine external annual grooves.

Thread-Cutting Tools

Thread-cutting tools have either triangular, square, or trapezoidal cutting edges, depending upon the cross section of the desired thread. Also, the plane angles of these tools must always be identical to those of the thread forms. Thread-cutting tools have straight shanks for external thread cutting and are of the bent-shank type when cutting internal threads.

Form Tools

Form tools have edges especially manufactured to take a certain form, which is opposite to the desired shape of the machined workpiece. A typical form tool usually includes the carbide tip, the chip breaker, the pad, the clamping screw (with a washer and a nut), and the shank. The function of the chip breaker is to break long chips every now and then, thus preventing the formation of very long twisted ribbons that may cause problems during the machining operation. The carbide tips (or ceramic tips) can have different shapes, depending upon the machining operations for which they are to be employed. The tips can either be solid or with a central through hole, depending upon whether brazing or mechanical clamping is employed for mounting the tip on the shank.

Unit 20　CNC Milling Machines

Text

There are two main types of milling machines, a vertical milling machine and a horizontal milling machine. The difference between the two is that the cutting tool used to cut the metal is either mounted horizontally or vertically.

Although automatic metalworking tools have been around in the United States since the early 1860's;[1] these machines were only capable of producing one component at a time and took a long time to configure. Towards the end of World War II and with the introduction of the jet engine, the need for more complex parts was increased, at the same time there was a period of conflict between machinists and management in large manufacturing companies. These factors combined to create the need for automatic machines that could produce large quantities of the desired components precisely, efficiently and in a reliable repetitive manner.

The first CNC milling machines were invented in the mid-20th century. John T. Parsons was the main contributor to the technology. He was heavily involved in making parts for airplanes and discovered that he could make more accurate parts using an IBM computer to move the cutting spindle in a smoother motion.[2] For about 100 years, all mills had been manual. While that was adequate, the invention of the CNC allowed for more accurate parts making.

CNC milling machines use a computer control to move and operate the spindle cutting head. The computer can move the head with absolute precision. The machines have a table where operators clamp down raw material to cut,[3] and many larger machines have automatic chip conveyors that move the metal chips away from the cutting area for collection.[4]

Most of CNC milling machines can perform versatile cutting and drilling operations. There are several types of these CNC machining centers that are well-known as multi-axis machining centers including 3-, 4- and 5-axis machines. Certainly, those machines which have the higher number of axes are cost higher than lower numbered axes because these machines are able to perform intricate parts production.[5] Thus, normally large scale manufacturers own 4-and 5-axis machines as both machine types are able to boost their productivity and profitability over a certain period of time.

CNC milling machines work by using a computer to send signals to a stepper motor controller. This then tells the stepper motor which direction to turn and how many steps to take. The motor is connected to the drive mechanism of the mill in the X, Y and Z axes. Some CNC milling machines use a servo motor instead of a stepper motor,[6] the advantage of this is that metal can be cut at high speeds and, due to a feedback loop, the machine always knows its exact position.

The main benefit of CNC milling is that the cutting process is controlled by a central computer. This eliminates a lot of the human errors which can exist with standard milling. Furthermore, whilst many machines are the less versatile vertical milling, the precise nature of computer control means that a lot of tasks impossible with human controlled vertical milling are possible with CNC vertical milling. The vertical cutting also can move along a Z-axis, allowing precise methods of cutting (such as engraving), not possible with manual machines. Although originally expensive, CNC milling has dropped in price thanks to the low price of computers and open source software. Most companies that offer machining offer CNC milling, meaning that if you approach a company for milling, they may be able to manufacture the whole widget for you.

Today many CNC milling machines are computer controlled vertical mills, and have the ability to move the spindle vertically along the Z axis. In CNC milling, end to end component design is automated using CAD/CAM programs. The program is put into the milling machine and the machine is then ready for production. Some machined components will generally require a number of different tooling applications such as drilling, reaming and tapping etc., and most modern machines will combine tools within a single cell. [7] This cell will move or rotate to apply the required tooling application, and this will also be controlled by the CNC system. With today's modern and complex machines, the machined part or workpiece can be moved from machine to machine automatically with the use of computer controlled robots, or human intervention, but in either case the steps needed to produce any part is highly automated and the finished part will closely match the CAD design.

Technical Words

component [kəmˈpəunənt]	n. 零件；组件；部件
engraving [inˈgreviŋ]	n. 雕刻；刻花
reaming [ˈrimiŋ]	n. 铰孔
robot [ˈrɔbət]	n. 机器人
automate [ˈɔːtəmeit]	v. （使）自动化
match [mætʃ]	v. 相配
versatile [ˈvəːsətai]	adj. 通用的，多才多艺的，万能的

Technical Phrases

vertical milling machine	立式铣床
horizontal milling machine	卧式铣床
rotary table	旋转台
clamp down	夹持
automatic chip conveyor	自动排屑器
metal chip	切屑
tool holder	刀架
tool change	换刀
multi-axis machining center	多轴加工中心
stepper motor	步进电动机
servo motor	伺服电动机
drive mechanism	驱动装置
feedback loop	反馈回路

end to end component design	点对点工件设计
CAD	计算机辅助设计
CAM	计算机辅助制造

Notes

(1) Although automatic metalworking tools have been around in the United States since the early 1860's.

文中 be around 的意思是"存在，出现"。

(2) He was heavily involved in making parts for airplanes and discovered that he could make more accurate parts using an IBM computer to move the cutting spindle in a smoother motion.

句中的 involved in making parts for airplanes 意思是"参与飞机零件的制造"；to move the cutting spindle in a smoother motion 可译为"使切削轴平稳运动"。

(3) The machines have a table *where operators clamp down raw material to cut* ...

句中 where operators clamp down raw material to cut 为定语从句，修饰 a table。

(4) ... many larger machines have automatic chip conveyors *that move the metal chips away from the cutting area for collection.*

句中 that move the metal chips away from the cutting area for collection 为定语从句，修饰 automatic chip conveyors；automatic chip conveyors 可译为"自动排屑器"。

(5) Certainly, those machines which have the higher number of axes are cost higher than lower numbered axes because these machines are able to perform intricate parts production.

句中出现 A is cost higher than B 的结构，意思是"A 比 B 贵"；higher number of axes 意思是"轴数较多"，lower numbered axes 意思是"轴数较少"。

(6) Some CNC milling machines use a servo motor instead of a stepper motor ...

句中 A instead of B 的结构表示选择，意思是"以 A 代 B"或"用 A 而不用 B"。

(7) Some machined components will generally require a number of different tooling applications such as drilling, reaming and tapping etc., and most modern machines will combine tools within a single cell.

根据具体内容，a single cell 可译为"刀库"。

Exercises

(1) Place a "T" after sentences that are true and an "F" after those that are false.

1) Generally speaking the vertical milling machine is more useful than the horizontal milling machine.

2) The first CNC milling machines were invented in the mid-20th century.

3) Machines which have the lower number of axes are cost higher than higher numbered axes.

4) All CNC milling machines work by using a computer to send signals to a servo motor controller.

5) CNC milling has dropped in price due to the low price of computers and open source software.

(2) Fill in the blanks according to the text with the words given below. Make changes if necessary.

vertical   intricate   intervention   horizontal   robot   versatile

There are two main types of milling machines, a _____ milling machine and a _____ milling machine. Most CNC milling machines can perform _____ cutting and drilling operations. Those machines which have the higher number of axes are able to perform _____ parts production. Nowadays, the machined part or workpiece can be moved from machine to machine automatically with the use of computer controlled _____, or human _____.

(3) Answer the following questions:
1) How many main types do CNC milling machines have? What are they?
2) When were the first CNC milling machines invented?
3) What is the first CNC milling machine made for manufacturing?
4) What do the CNC milling machines use to move and operate the spindle cutting head?
5) What is the function of the automatic chip conveyors?
6) What is the advantage of CNC milling machines using a servo motor instead of a stepper motor?

【参考译文】

第20课 CNC 铣床

铣床主要分立式铣床与卧式铣床两种，两者的差别在于切削材料用的刀具是水平安装或垂直安装。

尽管自19世纪60年代早期以来，美国就出现自动金属加工机床，但那些设备只能一次加工出一个工件，且准备工作花费的时间较长。到二次世界大战结束时，随着喷气发动机的问世，越来越需要加工更复杂的零件，与此同时许多大型制造公司的机械师与管理者之间曾一度存在冲突。这些因素综合起来产生了对自动机床的需求，以求精确、有效、可靠地重复加工所需要的大批零件。

第一台 CNC 铣床是20世纪中期研制出来的，John T. Parsons 对该项技术作出了主要贡献。他当时全身心地投入了飞机零件的加工制造，并从中发现用 IBM 计算机控制机床主轴平稳转动可以提高工件精度。100年来所有的铣床都是人工操作的，尽管这些铣床也够用了，但 CNC 铣床的发明可以加工出更精密的工件。

CNC 铣床是用计算机控制器驱动、控制主轴上的刀具。计算机可以使刀具的运动非常精确。这些机床配有工作台，操作者可以夹持原材料进行切削；一些大型机床还装有自动排屑器，以便清除加工区的切屑并收集起来。

大多数 CNC 铣床可以进行各种切削和钻削操作。CNC 加工中心的种类很多，但众所熟知的要算多轴加工中心，其中包括3轴、4轴、5轴机床。轴数越多肯定价格越贵，因为这些机床能够完成复杂零件加工生产。因此，大型制造厂家通常都拥有4轴和5轴机床，因为这两种机床都能够在一定时期内大幅提升生产率和盈利率。

CNC 铣床是这样工作的：计算机将信号传输给步进电动机控制器，然后控制器告知电动机旋转方向以及转动多少步。步进电动机与铣床 X、Y、Z 轴的传动装置相连。有些 CNC 铣床使用伺服电动机而不是步进电动机，其优点是切削速度高，其次由于存在反馈回路，机床总能知道其确切位置。

CNC 铣床的优势在于切削过程由中央计算机控制，从而消除了标准铣床存在的许多人工错误。其次，虽然许多立式铣床的通用性差一些，但计算机控制的精确性好，这意味着原来人工操控的立式铣床无法完成的任务可用 CNC 立式铣床完成。此外，立式机床可沿 Z 轴运动，能够实现一些精密切削（如雕刻），而这些加工则是人工操控无法实现的。尽管最初价格昂贵，但由于计算机成本降低、软件开放，CNC 铣床价格也随之降低。大多数机床销售商也都经营 CNC 铣床，也就是说，只要你联系铣床销售商，他们就能够为你提供全套设备。

当今许多 CNC 铣床是计算机控制的立式铣床，其主轴可沿 Z 轴垂直运动。在 CNC 铣床中，用 CAD/CAM 软件可以实现点对点零件设计。只要将程序输入到铣床，机床随时可以加工。有些工件通常需要进行铣削、铰孔、攻螺纹等一系列加工，现代机床则在刀库中集中了所需刀具；刀库可以移动、旋转，以适应所需要的加工方法，刀库也是用 CNC 系统控制的。对于现代高端机床来说，工件既可以用计算机控制的机器人从这个机床自动地输送到另一台机床，也可以用人工干预的方式完成；但不管哪一种情形，加工零件所需要的步骤都高度自动化，成品都将更接近于 CAD 设计要求。

【Reading Material】

Benefits of CNC Milling Machines

The CNC milling machine is one of the most important CNC machine tools which are widely used in many industrial businesses. It has been effectively used to increase one's profitability as well as productivity in performing repetitive high precision and difficult milling operations including drilling, slot and keyway cutting and many more, which could not be possibly done manually over a long period of time. Therefore, these CNC milling machines are a compulsory tool in most industrial businesses which have succeeded the traditional milling methods.

The invention of the CNC milling machine had a large impact on the manufacturing sector, as the machine could cut parts over and over again with absolute accuracy. Once the program is inputted into the machine, as long as the tools are sharp, the machine will cut every piece exactly the same.

CNC milling machines are set up by trained machinists and can be run by less-skilled operators. Operators place raw material in the machine when parts are done and monitor the machine for tool breaks and errors. This frees up the machinist to take on other tasks in the machine shop. Previously, a machinist was making parts by hand and was seldom able to multitask.

Basically, CNC milling machine operations are controlled by its own distinctive control software. Operators need to be proficiency in CAD and CAM programming skills in order to control the computerized machining centers when performing high precision milling operations. As compared with traditional milling machines (or manual machines), most CNC machining centers are able to perform high precision milling operations and to reduce unnecessary wastage.

As mentioned earlier, most CNC machines are capable of fabricating components without sacrificing precision and quality. Thus, using such CNC machines can effectively reduce one's overhead costs.

Unit 21　Spindles of CNC Machine Tools

Text

Machine Tool Spindles

The spindle is a very important and integrated component of different machine tools. Though the spindles are used in different fields and for different purposes, those which are used in machine tools are called machine tool spindles. They are rotating devices used to hold and act as axes for cutting tools or workpieces on lathes, milling machine and different other machine tools.

Rotational Accuracy of Spindles

At the high cutting speeds and high material removal rates, the spindles carrying the workpiece or the tool are subject to deflection and thrust forces. [1] To ensure increased stability and minimize torsional strain, the machine spindle is designed to be short and stiff and the final drive to the spindle is located as near to the front bearing as possible. [2] The rotational accuracy of the spindle is dependent on the quality and design of bearing used. The ball or roller bearings are suitable for high speeds and high loads because of low friction, lower wear rate and lesser liability to incorrect adjustment and ease of replacement when necessary. For efficient service and accuracy the bearings should be of high quality. The vibrations and noise in the spindle can be reduced by using toothed belts and accurate and balanced gears. Adequate supply of lubricants should be ensured to the spindle bearings.

Categories of Spindles on CNC Machines

At today's fast age of growth and technology, the use of CNC spindles by many manufactures is increasingly on the rise. The use of the right spindle in a CNC machine can make all the difference in terms of productivity. [3] The spindle used on the CNC machine can make any tooling application whether milling, grinding, or metal finishing fast and easy.

Machine tool spindles (See Fig. 21-1) are basically rotating devices so some type of medium is required to drive these spindles. These drives may be belt, motor, gear, shaft or any other medium.

Fig. 21-1　CNC Machine Tool Spindles

- **Belt Driven Spindles**: These spindles have a pulley at the end to drive via a belt.
- **Motorized Spindles**: These spindles are provided with a drive motor integrally built into the spindle.
- **Gear Driven Spindles**: These spindles are provided with gear coupling to allow for high or low-speed spindle rotation.
- **High Frequency Spindles**: These spindles are equipped with separate external frequency converter used to drive them at very high speed.

In addition there are other classifications of spindles used on CNC machines depending on the uses and applications. Some common types are as follows:
- CNC High Frequency Motor Spindle
- CNC Machine Center Spindle
- CNC Surface Grinding Spindle
- CNC Turning Machine Spindle
- CNC External Grinding Spindle
- CNC Special Purpose Machine Spindle

Proper Selection of Spindles

Proper selection of a spindle used in a CNC machine is imperative for optimum machine performance. The electric spindle is the heart of the CNC machine. Many factors must be considered when selecting the correct spindle like:
- Material to be cut
- Tooling
- Production volume
- Machine feed rate
- Spindle rpm

Generally, each material to be cut has an ideal tool profile and cutting speed. Larger diameter tools perform in slower speeds while smaller diameter tools perform in higher speeds. For any given cut, spindle speed and feed rate must be balanced to get the best quality, tool life and spindle life. Incorrect spindle speed is a common mistake in all types of CNC machining. Machine feed rate and spindle rpm are directly related to one another. For faster machine cutting, the higher should be the rpm. Typically, feed rates that are very slow will lead to decrease in tool life because of increased friction. Other considerations are like how to enter the part. Ramping into the part is the most easy and preferred method. The ideal ramp should be anywhere between 0 and 20 degrees from the table surface. This angle will allow the material to enter at 100% of the feed rate. At any angle greater than 20 degrees, the feed rate should be reduced. Entering into the part on a ramp will increase spindle bearing and tool life greatly.

Technical Words

bearing ['bɛəriŋ]	n. 轴承
ball [bɔːl]	n. 滚珠
friction ['frikʃən]	n. 摩擦；阻力
noise [nɔiz]	n. 噪声，干扰
lubricant ['lubrikənt]	n. 润滑油，润滑剂
technology [tek'nɔlədʒi]	n. 工艺；技术
motor ['məutə]	n. 电动机；原动力
frequency ['frikwənsi]	n. 频率；频繁
torsional ['tɔːʃənl]	adj. 扭力的；扭转的

strain [strein]   n. 张力，应变
             v. 拉紧，尽力

Technical Phrases

rotating device	旋转设备
thrust force	推力
torsional strain	扭转应变
cutting speed	切削速度
roller bearing	滚珠轴承
wear rate	磨损率
spindle bearing	主轴轴承
gear coupling	齿轮联轴器
frequency converter	频率变换器
feed rate	进给率
tool profile	刀具外形，刀具轮廓

Notes

(1) At the high cutting speeds and high material removal rates, the spindles carrying the workpiece or the tool are subject to deflection and thrust forces.

文中 material removal rate 是"材料切削率"的意思。subject to... 可译为"易受……影响"。

(2) To ensure increased stability and minimize torsional strain, the machine spindle is designed to be short and stiff and the final drive to the spindle is located as near to the front bearing as possible.

句中 as near to the front bearing as possible 可译为"尽可能靠近前轴承"。

(3) The use of the right spindle in a CNC machine can make all the difference in terms of productivity.

文中 make all the difference 在这里有"作用大不相同"的意思；in terms of 的意思是"从……方面来说，从……角度来说"。

Exercises

(1) Place a "T" after sentences that are true and an "F" after those that are false.

1) Spindles are rotating devices used to hold and act as axes for cutting tools or workpieces.

2) The machine spindle is designed to be long and stiff and the final drive to the spindle is located as far to the front bearing as possible.

3) The vibrations and noise in the spindle can be increased by using toothed belts and accurate and balanced gears.

4) Some type of medium is required to drive machine tool spindles.

5) Larger diameter tools perform in higher speeds while smaller diameter tools perform in slo-

wer speeds.

(2) Fill in the blanks according to the text with the words given below. Make changes if necessary.

rotational   vibration   imperative   toothed   balanced   deflection

During cutting, spindles are subject to _____ and thrust forces. The _____ accuracy of the spindle is dependent on the quality and design of bearing used. The _____ and noise in the spindle can be reduced by using _____ belts and accurate and balanced gears. Proper selection of a spindle used in a CNC machine is _____ for optimum machine performance. For any given cut, spindle speed and feed rate must be _____ to get the best quality, tool life and spindle life.

(3) Answer the following questions:
1) What is the function of spindles?
2) What are the spindles subject to during cutting parts?
3) What is the rotational accuracy of the spindle dependent on?
4) Which kind of media may be required to drive spindles?
5) How many factors must be considered when selecting the correct spindle? What are they?
6) Why do very slow feed rates lead to decrease in tool life?

【参考译文】

第 21 课 数控机床主轴

机床主轴

主轴是各种机床中很重要的集成组件。尽管主轴用于不同场合，也有不同用途，但机床上所用的主轴被称之为机床主轴。主轴是旋转设备，用于支撑车床、铣床以及其他机床上的刀具或工件，或作为刀具和工具的转轴。

主轴旋转精度

在切削速度和材料切除率较高的情形下，携带工件或刀具的主轴容易发生偏斜并受到推力的作用。为了确保提高稳定性并把扭转应变降到最低，机床主轴设计得短且具备刚性，主轴的终端驱动装置安装得尽可能靠近前轴承。主轴的旋转精度取决于所用轴承的设计和质量。由于摩擦小、磨损率低、受校准不当的影响小、易于替换等缘故，滚珠轴承、滚柱轴承适合于高速切削及重载加工。为提高效率和加工精度，应该使用优质轴承。使用齿形带以及精确而配平的齿轮，可以降低主轴振动和噪声。此外还应保证为主轴轴承提供充足的润滑油。

数控机床主轴分类

在技术快速发展的今天，许多制造厂家越来越多地使用数控主轴。使用合适的主轴可以使数控机床的生产效率大幅提升。数控机床所用的主轴可使铣削、磨削、金属表面精整等任何加工快速而便捷。

机床主轴（见图 21-1）基本上是旋转设备，因而需要某种装置来驱动。这些驱动装置有可能是传动带、电动机、齿轮、轴体或其他装置。

- **带驱动式主轴**：此类主轴末端装有带轮，通过带传动来驱动。
- **电动机驱动式主轴**：此类主轴靠内置驱动电动机驱动。
- **齿轮传动式主轴**：此类主轴通过齿轮连接使主轴高速或低速转动。
- **高频主轴**：此类主轴配有独立的外置变频器，驱动主轴高速旋转。

除此以外，数控机床使用的主轴还可以按用途分类，常见类型有以下几种：
- 数控高频电主轴
- 数控加工中心主轴
- 数控平面磨削主轴
- 数控车削机床主轴
- 数控外圆磨削主轴
- 数控专用机床主轴

正确选择主轴

正确选择加工主轴对于数控机床发挥其最佳性能是必不可少的。电主轴是数控机床的核心部件，选择主轴需要考虑许多因素，如：
- 切削的材料
- 使用的刀具
- 生产的规模
- 机床进给率
- 主轴的转速

一般来说，被切削的每种材料有其理想的刀具轮廓和切削速度。大直径刀具低速切削，而小直径刀具则高速切削。对于给定的切削加工，主轴转速与进给率必须匹配才能获得最佳切削质量、刀具寿命和主轴寿命。主轴转速不合适是所有数控加工方式中常见的错误。机床进给率与主轴转速之间有直接的联系。对高速切削而言，主轴转速应该高一些。通常，进给率过低，摩擦就大，导致刀具寿命缩短。其他应考虑的因素之一是如何使刀具切入工件。刀具倾斜切入是最容易也是最佳方法。理想的倾斜度应当是从工作台表面算起0°~20°之间的任一角度。这个角度可使材料100%进给率进入。切入角大于20°时，应该降低进给率。刀具倾斜切入工件会大大提高主轴轴承和刀具的使用寿命。

【Reading Material】

Different Spindles of CNC Machine Tools

Most machine tool spindles are driven by motors, belts, gears, hydraulic or pneumatic and other ways. They are found in different configurations suiting variety of machines and applications. Some machine tool spindles are encased in solid boxes or box like cases. Some others are bolted at the bottom of housing. Sometimes multiple spindle heads are used to speed up the processing and machining operations. To drive these spindles different types of bearings are used with the machines such as hydrostatic, roller, air, angular contact etc.

The criterion, for defining types of machine tool spindles, covers a wide range. We can define the types on the basis of applications, performance specifications, medium of drive, configuration etc.

Machine Tool Spindle Based on Applications

Machine tool spindle has found variety of applications in a number of fields such as drilling, boring, milling, turning, grinding etc.

Drilling spindles: They are designed for drilling operations and have good thrust capacity in addition to radial load rating.

Boring spindles: They are designed for boring operations, used in the machining of internal diameters of workpieces.

Milling spindles: They are designed for various milling operations and are used with a wide range of machine tools.

Turning spindles: They are designed for working in vertical or horizontal lathes or other machinery tools.

Grinding spindles: They are designed for working in grinding wheels for precision, size, and surface finishing.

Machine Tool Spindle Based on Performance Specifications

A well recognized criterion used for specifying spindles is their performance. It includes spindle power, operating speed, input voltage, maximum torque etc.

Machine tool spindle: A rotating component is used to hold the cutting tools that shape a work piece. Modern CNC machine centers are configured with a cartridge style spindle. We call this a standard spindle and it is the most common spindle repair.

High speed spindle: A spindle that can operate at speeds above 10,000 revolutions per minute. This type of spindle is typically belt driven.

Inline Spindle: A spindle that is directly coupled to an inline motor. Inline spindles can typically operate at speeds of up to 30,000 revolutions per minute.

Unit 22　Computer Graphics Programming

Text

The newest form of NC programming is called CAM. CAM stands for computer aided manufacturing. When using a CAM system, the programmer either calls up an existing part drawing, or defines the part geometry to the computer. Next, the cutter path is drawn around the part and the necessary information on cut direction, tool, speeds, and feeds is input. This information is then converted into either an APT file or a cutter centerline data file by a postprocessor. This data is then fed through a secondary postprocessor, which produces the necessary NC code for a given machine.

There are a number of different CAM systems on the market. The following is a brief explanation of several different types.

Digitizing Systems

Digitizing systems use an existing part drawing to obtain the geometry information. The drawing used must be drawn to a true scale of the finished part. The scale drawing is fed into a digitizer which is connected to the CAM system computer. A digitizer is a device consisting of a table with a probe or other sensor attached. The sensing device is passed over the drawing, converting drawing lines in the necessary mathematical information into electronic form which the computer needs to recreate the drawing. [1] The cutter path is then defined by the programmer and the information of the cutter, speed, and feed is input. The result is then postprocessed into the necessary tape code for the CNC machine.

Digitizing is one of the simplest CAM methods, but it is also the least accurate. The accuracy of the part geometry is dependent upon the accuracy of the scaled drawing which was digitized. Fortunately this shortcoming is minimized by the fact that drawings 30 times size or larger can be used.

Scanning Systems

Scanning systems use the part itself rather than a part drawing to obtain the geometric database to machine the part. Scanning is used when complex curves are to be machined which are difficult to draw and which do not fit a true mathematical model. [2] Automobile bodies are an example of such curves. A scanner is a probing device connected to the CAM system computer which is passed over a model of the part. [3] The probe feeds information concerning the part geometry into the computer which then calculates the points necessary to define the part shape. [4] The cutter path, tool data, speeds, and feeds are then input. The information is then fed to the postprocessor which will convert the information into the necessary tape coding.

CAM Systems

Although the two methods mentioned above are referred to as CAM systems, CAM is also the term used to describe computer graphics programming. CAM systems can be run on a mainframe computer, or on a PC. Mainframe systems are found in larger plants primarily involved with four- or five-axis programming. Microcomputer-based CAM systems function well for three-axis. New systems can handle four-axis programming as well. Microcomputer systems are an economical choice for many small to midsize shops.

In CAM (graphics) programming, the programmer defines the part geometry to the computer using one or more of several input devices. These devices may be the keyboard, a mouse, a digitizer, or a light pen. After the part geometry is defined, the cutter path is drawn around the part. Information on the cut direction, tool data, speeds and feeds are then input. This information is then translated by a series of postprocessors into the necessary NC code.

Technical Words

geometry [dʒiˈɔmitri]	n. 几何学
postprocessor [pəustˈprəusesə]	n. 后置处理器
probe [prəub]	n. 探针
	v. 探测
mouse [maus]	n. 鼠标
database [ˈdeitəbeis]	n. 数据库
digitizer [ˈdidʒitaizə]	n. 数字转换器

Technical Phrases

secondary postprocessor	次后置处理器
scanning system	扫描系统

Notes

(1) The sensing device is passed over the drawing, converting drawing lines in the necessary mathematical information into electronic form *which the computer needs to recreate the drawing.*

which 引导的定语从句修饰 electronic form；pass over 意为"扫描"。

(2) Scanning is used when complex curves are to be machined *which are difficult to draw and which do not fit a true mathematical model.*

主句为：Scanning is used. which are difficult to draw 和 which do not fit a true mathematical model 是修饰 complex curves 的两个定语从句。

(3) A scanner is a probing device connected to the CAM system computer *which is passed over a model of the part.*

which 引导定语从句，修饰 the CAM system computer。

(4) The probe feeds information concerning the part geometry into the computer *which then calculates the points necessary to define the part shape.*

which 引导定语从句，修饰 computer。

Exercises

(1) Place a "T" after sentences that are true and an "F" after those that are false.

1) According to the first paragraph, we conclude that there are two postprocessors in a CAM

system.

2) Digitizing is the simplest but the least accurate method.

3) Scanning systems use a probe to obtain information concerning the part and feed that information into the computer.

4) From the last two paragraphs, we conclude the term CAM means computer aided manufacturing.

(2) Fill in the blanks according to the text with the words given below. Make changes if necessary.

  part translate digitizing complex geometric

There are many CAM systems on the market. The most commonly used types are digitizing, scanning, and CAM systems. _____ system is one of the simplest types. It uses an existing part drawing to obtain the _____ information. As a comparison, scanning systems use the _____ itself to obtain the geometric database. This system can be used for machining _____ curves such as automobile bodies. After the part geometry is defined, all the necessary information must be _____ into the necessary NC codes.

(3) Answer the following questions:

1) What are the advantages and disadvantages of digitizing systems?

2) What is the function of the sensing device of the digitizing system?

3) Give some examples of input devices.

【参考译文】

第22课 计算机图形编程

  CAM 是指计算机辅助制造，这是数控编程的最新形式。使用 CAM 系统时，编程员首先调用现存的零件图，或给计算机输入零件几何参数。其次，CAM 系统绘制加工时的走刀路径，操作员输入切削方向、刀具、转速及进给量等必要信息。然后，后置处理器把这些信息转化成 APT 文件或刀具中心线数据文件。将这些数据送入次级后置处理器，并生成所选机床需要的数控代码。

  市场上有许多 CAM 系统。现就几种不同类型 CAM 系统简要介绍如下。

数字化系统

  数字化系统可从按比例缩放过的零件图中获得几何数据。使用时，先将零件图输入到与 CAM 计算机联起来的数字转换器。数字转换器由工作台和与之相联的探针或其他传感器组成。其次，传感装置扫描样图，把数学形式的样图轮廓信息转换成电量形式，这是计算机重建样图必需的。然后，编程人员定义走刀路径，输入刀具、转速、进给量参数。最后，经后置处理便生成 CNC 机床需要的纸带代码。

  数字化方法是最简单的 CAM 方法之一，但也是精确度最差的方法，零件的几何精度取决于数字化比例绘图的精度。所幸的是，若使用30倍甚至更大的样图，精度上的缺陷就显得微不足道了。

扫描系统

扫描系统使用零件本身而不是零件样图来建立加工需要的几何参数数据库。在加工一些复杂曲线时就要使用扫描法，因为这些曲线很难绘制，且难以建立合适的数学模型，汽车车身就是一个例子。扫描仪是一种与 CAM 系统计算机联起来的探测装置，使用时由探针逐点扫描零件模型，并把零件几何参数信息输入计算机，然后计算机算出确定零件轮廓所需要的一些点，再输入走刀路径、刀具、速度及进给量参数，接着信息被输进后置处理器，后置处理器把信息转变成所需要的纸带代码。

CAM 系统

上述两种方法虽然都称做 CAM 系统，但 CAM 一词也是描述计算机图形编程的术语。CAM 系统可在主机上运行，亦可运用于个人计算机。主机系统多见于大型工厂，且主要针对 4 轴或 5 轴编程。基于微型计算机的 CAM 系统则更适合于 3 轴编程，但一些新版 CAM 系统也能处理 4 轴编程。微型计算机 CAM 系统对很多中小型工厂来说较为实用。

在 CAM 图形编程中，编程员可以使用一种或多种输入装置给计算机输入零件几何参数，这些装置可以是键盘、鼠标、数字转换器和光笔。设好零件几何参数后，进给路径被绘出来，然后再输入切削方向、刀具、转速及进给量参数，这些参数再被一些后置处理器转换成所需要的数控代码。

Unit 23　CAD

Text

The advent of the computer proved to be a boon to the design engineer in that it simplified the long, tedious calculations which were often involved in designing a part. [1] In 1963, the Massachusetts Institute of Technology (MIT) demonstrated a computer system called Sketchpad that created and displayed graphic information on a cathode-ray tube (CRT) screen. This system soon became known as CAD, and it allows the designer or engineer to produce finished engineering drawings from simple pencil sketches or from models and modify these drawings on the screen if they do not seem functional. [2] From three-drawing orthographic views, the designer can transform drawings into a three-dimensional view and, with the proper computer software, show how the part would function in use. This enables a designer to redesign a part on the screen, project how the part will operate in use, and make successive design changes in a matter of minutes.

CAD Components

CAD is a televisionlike system that produces a picture on the CRT screen from electronic signals received from a computer. [3] Most CAD systems consist of a desk-top computer which is connected to the main or host computer. The addition of a keyboard, light pen, or an electronic tablet and plotter enables the operator to produce any drawing or view required.

The operator generally starts with a pencil sketch and, with the use of the light pen or an electronic tablet, can produce a properly scaled drawing of the part on the CRT screen and also record it in the computer memory. [4] If design changes are necessary, the designer is able to create and change parts and lines on the CRT screen with a light pen, an electromechanical cursor, or an electronic tablet.

After the design is finished, the engineer or designer can test the anticipated performance of the part. Should any design changes be necessary, the engineer or designer can make changes quickly and easily to any part of the drawing or design without having to redraw the original. [5] Once the design is considered correct, the plotter can be directed to produce a finished drawing of the part.

Advantages of CAD

CAD offers industry many advantages which result in more accurate work and greater productivity. [6] The following is a list of some of the more common CAD advantages:
- Greater productivity of drafting personnel
- Less drawing production time
- Better drawing revision procedures
- Greater drawing and design accuracy
- Greater detail in layouts
- Better drawing appearance
- Greater parts standardization

- Better factory assembly procedures
- Less scrap produced

Technical Words

advent [ˈædvənt]	n. 出现，到来
boon [buːn]	n. 实惠
orthographic [ˌɔːθəˈɡræfik]	adj. 正字法的；正交的；投影的
cursor [ˈkəːsə]	n. 指针，光标
layout [ˈleiˌaut]	n. 布局；设计；规划
scrap [skræp]	n. 废料

Notes

(1) The advent of the computer proved to be a boon to the design engineer in that it simplified the long, tedious calculations *which were often involved in designing a part.*

which 引导的定语从句修饰 calculations；in that 作"因为"讲。

(2) This system soon became known as CAD, **and** it allows the designer or engineer to produce finished engineering drawings from simple pencil sketches or from models and modify these drawings on the screen if they do not seem functional.

本句由两个子句组成，且以 **and** 相连；第二个子句的主语 it 指 CAD。动词不定式 to produce… 和（to）modify… 作宾补。

(3) CAD is a televisionlike system *that produces a picture on the CRT screen from electronic signals received from a computer.*

that 引导的定语从句修饰 a televisionlike system。过去分词短语 received from a computer 作定语修饰 electronic signals。

(4) The operator generally **starts** with a pencil sketch and, with the use of the light pen or an electronic tablet, **can produce** a properly scaled drawing of the part on the CRT screen and also **record** it in the computer memory.

本句主语是 The operator，谓语有三个：**starts，can produce，（can）record**；介词短语 with… 作状语。

(5) Should any design changes be necessary, the engineer or designer can make changes quickly and easily to any part of the drawing or design without having to redraw the original.

本句中 Should any design changes be necessary 是倒装句，正常语序为：if any design changes should be necessary。

(6) CAD offers industry many advantages *which result in more accurate work and greater productivity.*

which 引导的定语从句修饰 advantages。

Exercises

(1) Place a "T" after sentences that are true and an "F" after those that are false.

1) Designing a part used to involve long tedious calculations.

2) Using CAD system, a designer can design and modify a part drawing and see how it will operate in use on the screen.

3) CAD is a system that can use a pencil and paper to produce a part drawing quickly.

4) If design changes are necessary, the designer has to redraw the part drawing completely on the CRT screen.

5) Using CAD system, industry can produce more accurate work and obtain greater productivity.

(2) Fill in the blanks according to the text with the words given below. Make changes if necessary.

  memory plotter screen modify signal

CAD is a computer system which creates and displays graphic information on a cathode-ray tube _____. It allows the designer to produce engineering drawings and _____ them on the screen if they do not seem functional. CAD produces a picture on the CRT screen from electronic _____ from a computer. The operator can use CAD system to produce a proper scaled drawing of the part on the CRT screen and record it in the computer _____. If the design is correct, the _____ can be directed to produce a finished drawing of the part.

(3) Answer the following questions:

1) Why is the advent of computers considered to be a good thing to design engineers?

2) What devices can be used by the operator to input the necessary information to produce the drawing?

3) Why is CAD widely used in industry?

【参考译文】

第 23 课　计算机辅助设计

  计算机对设计师非常实用，因为它简化了零件设计中冗长而乏味的计算。1963 年，美国麻省理工学院曾经演示了称为 Sketchpad（画图板）的计算机系统，它能够在阴极射线管（CRT）显示屏上生成并显示图形信息。该系统很快成为所谓的 CAD，这使设计者根据草图或模型便可绘制最终的工程图样；如果不符合要求，还可在显示屏上即时修改。此外，从三视图投影的角度，设计者能把视图转变成为三维图形；使用合适软件，还可模拟零件的实际使用情况。这样，设计者可在显示屏上重新设计零件图、观察零件使用情况、不断改进图样，从而缩短了设计周期。

  CAD 的组成

  CAD 类似于电视系统，它接收计算机发来的电信号，在 CRT 显示屏上生成图像。多数 CAD 系统使用台式计算机，并与主机相联。配以键盘、光笔、电子板、绘图仪等附件，操作者可以绘制任何零件图。

  只要有了草图，操作者使用光笔或电子板通常就能在 CRT 显示屏上绘制适当比例的零件图，并储存于计算机内存之中。如果需要改进，设计者可用光笔、机电光标、电子板在 CRT 显示屏上直接修改，便可生成新的零件图。

完成设计后，工程师或设计师可以检验零件是否达到预期性能。若需要改进，工程师或设计师可迅速、方便地对设计图的任何部分进行修改而无需重绘原图。一旦确认设计结果正确无误，便可用绘图仪直接打印最终的零件图样。

CAD 的优点

CAD 给工业带来许多实惠，使绘图工作高效而精确。下面列出 CAD 的一些常见优点：
- 提高绘图效率
- 缩短绘图时间
- 便于图样修改
- 改善绘图与设计精度
- 细节规划好
- 图样外观质量好
- 提高零件标准化程度
- 装配工艺合理
- 废图少

Unit 24　CAM

Text

CAM started with NC in 1949 at MIT. This project, sponsored by the U. S. Air Force, was the first application of computer technology to control the operation of a milling machine. [1]

Standard NC machines greatly reduced the machining time required to produce a part or complete a production run of parts, but the overall operation was still time-consuming. Tape had to be prepared for the part, editing the program would result in making a new tape, and tapes had to be rewound each time a part was completed. With this in mind, the machine manufacturers added a computer to the existing NC machine, introducing the beginning of CNC.

The addition of the computer greatly increased the flexibility of the machine tool. The parts program was now run from the computer's memory instead of from a tape that had to be rewound. [2] Any revisions or editing of the program could be done at the machine, and changes could be stored.

As the machine tool manufacturers continued to improve the efficiency of their machines, the computer capabilities were greatly increased to programmable microprocessors, and many time-saving devices were introduced to increase the machine's cutting time and reduce downtime. Some of these machine options are automatic tool changers, parts loaders and unloaders, chip conveyors, tool wear monitors, inprocess gaging and robots—which brings us to today's machining centers (Fig. 24-1). [3]

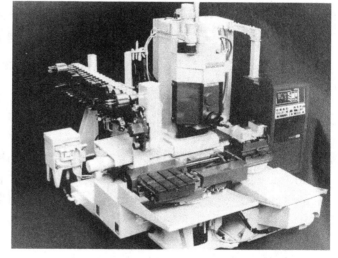

Fig. 24-1　Machining centers with automatic tool changers and workhanding equipment

CAM uses all the advanced technologies to automate the operations in manufacturing and handle the data that drives the process. The tools of CAM include computer technologies, CAE, and robotics. CAM uses all these technologies to join the process of design with automated production machine tools, material handling equipment, and control systems. Without computers, the most important tool in industry, the productivity of the United States would be in serious trouble. Computers help people to become more productive and to do things that would almost be impossible without them.

CAM ties together all the major functions of a factory. The manufacturing or production operations are joined together with the process planning, production scheduling, material handling, inventory control, product inspection, machinery control, and maintenance to form a total manufacturing system.

A CAM system generally contains three major divisions:

Manufacturing: The physical operation of controlling the machine tools, material handling equipment, inspection operations, etc., in order to produce the parts required.

Engineering: The process which involves design and engineering activities to ensure that the parts are designed properly in order to function as required.

Management: The information such as scheduling, inventory control, labor, and manufacturing costs, and all the data required to control the entire plant.

CAM increases the productivity and versatility of machine tools. Before the introduction of NC and CAM, most machine tools were cutting metal only about 5 percent of the time. The automated systems available now cut metal about 70 percent of the time, and the goal is to come as close as possible to having them remove metal 100 percent of the available time.[4]

Technical Words

automate ['ɔːtəmeit]	v. 使自动化
inventory ['invəntri]	n. 详细目录，财产清册，报表
versatility [ˌvɜːsə'tiləti]	n. 多功能性，多样性
rewind [riː'waind]	v. 重绕

Technical Phrases

production run	生产过程
time-consuming	费时
time-saving	省时
automatic tool changer	自动换刀装置
part loader and unloader	零件装卸装置
chip conveyor	排屑器

Notes

(1) This project, sponsored by the U.S. Air Force, was the first application of computer technology to control the operation of a milling machine.

句中 sponsored by the U.S. Air Force 是插入语，进一步说明 This project。

(2) The parts program was now run from the computer's memory instead of from a tape *that had to be rewound.*

instead of 把介词短语 from the computer's memory 和 from a tape 连接起来，表示前者取代后者；that 引导的定语从句修饰 a tape。

(3) Some of these machine options are automatic tool changers, parts loaders and unloaders, chip conveyors, tool wear monitors, in-process gaging and robots — which brings us to today's machining centers.

which 指代的是上面所提的所有事情，把它们看成一个整体，故谓语动词用单数。

(4) The automated systems available now cut metal about 70 percent of the time, and the goal is to come as close as possible to having them remove metal 100 percent of the available time.

available 一般放在被修饰词之后；as close as possible 意为"与……尽可能接近"。

Exercises

(1) Place a "T" after sentences that are true and an "F" after those that are false.

1) CAM system was first used to control the operation of a milling machine.
2) NC machines are more time-consuming than CNC machines in terms of overall operation.
3) CAM uses automatic tool changers to automate the operations in manufacturing in order to increase the machine's cutting time.
4) The application of computer to controlling a machine tool greatly increases its flexibility but reduces the productivity.

(2) Fill in the blanks according to the text with the words given below. Make changes if necessary.

  advance productivity exist automate standard changer memory

The overall operation of a(an) _____ NC machine was still time-consuming. It used tape to store programs, so a computer was added to the _____ NC machine, and CNC appeared. In a CNC machine, a program is stored in the computer's _____. Many time-saving devices such as automatic tool _____ were introduced to increase the machine's cutting time and reduce downtime. CAM uses all the _____ technologies to _____ the operations in manufacturing and handles the data that drive the process. So it can greatly increase the _____.

(3) Answer the following questions:

1) Why is standard NC still time-consuming?
2) Where is the part program stored when using a CAM system?
3) Which kind of device was introduced to the machine to increase machine's cutting time? Can you give some examples?
4) How many divisions does a CAM system contain? What are they?

【参考译文】

第 24 课 计算机辅助制造

  随着数控技术的发展，计算机辅助制造（CAM）于 1949 年在美国麻省理工学院问世，这一项由美国空军资助的项目是计算机控制技术在控制铣床中的首例应用。

  标准 NC 机床的运用虽然极大地缩短了生产单个零件或完成一批零件整个生产过程需要的时间，但整体运作过程仍然相当费时。这是由于必须为每种零件准备纸带、每次编辑程序都要制作新的纸带、每加工完一个零件后还要卷带的缘故。鉴于这些因素，机床生产厂家在现有 NC 机床中引进了计算机，于是出现了计算机数字控制（CNC）技术。

  计算机的引入极大地增强了机床柔性。现在，零件加工程序存于计算机存储器中，而不是存于总要卷带的穿孔带上；程序的修改与编辑也在机床上进行，并能存储修改结果。

机床效率持续被生产厂商提高，要求计算机内存容量越来越大，可编程微处理器便应运而生。由于机床中加入了微处理器及其他一些高效省时设备，从而增加了切削时间，缩短了停机时间。有些机床还具有自动换刀、自动装卸工件、自动排屑、自动监控刀具磨损、在程检测、机器人等备选功能——所有这些造就了现代加工中心（见图24-1）。

CAM使用所有先进技术来提升制造过程的自动化程度和方便加工过程的数据处理。这些CAM技术包括：计算机技术、计算机辅助工程（CAE）以及机器人技术。CAM系统运用这些技术把设计过程与自动加工机床、物料处理设备及控制系统集为一体。假如没有计算机这一工业中最重要的工具，美国的生产效率就会陷入严重困境。计算机不仅有助于人们提高工作效率，而且帮助人们完成许多没有计算机就几乎无法做的事情。

CAM甚至把工厂的所有主要功能都联系起来了。它把制造或生产运营与工艺设计、生产计划、物料处理、报表管理、产品检验、机械控制及维修联在一起，从而形成一个完整的制造系统。

CAM系统一般分为三个主要部分：

制造：完成对机床、物料处理、运行检验等部分的操作控制，以加工所需零件。

工程：管理设计与工艺活动，确保零件设计合理，实现预期的功能要求。

管理：指管理计划、报表、劳务、制造成本方面的信息，管理整个工厂所需要的全部数据。

CAM提高了机床的生产效率，强化了机床加工功能的多样性。在NC与CAM出现以前，大多数机床只有5%的时间用于加工，而现代自动化加工系统的加工时间占到70%，其最终目标是把加工时间尽可能提高到100%。

Unit 25 CAD/CAM/CNC

Text

CAD/CAM systems can be used to produce CNC data to machine a part. After preparing a tool list and setup plan for the required part, the CNC programmer starts by creating a database. Once this database has been created, the programmer can recall the part on the CRT screen. After the part is displayed, the programmer describes the tools required from the information in the tool library. This library contains a description and either a name or a tool identification number for each and every tool available for use. Assume that the database and tools described are for a milling/drilling operation. The next operation would then be to generate the tool path. In this case, the types of machining that could be performed would fall into three categories:

Drilling operations: Using the Z axis perpendicular with the surface of the part

Profile milling operations: Milling the profile of the part

Pocket milling operations: Using the Z axis to plunge and remove the material from the interior of the part

The CRT screen displays a graphic representation of the part and the path the cutting tools will follow in order to complete the machining of the part. This information must now be converted to a cutter location (CL) file. With a CL file, the information that was created to generate the graphic display is converted into coordinate locations to move the cutting tool around the part.[1] This information is now in a readable format.

Since a large variety of machines and machine controllers are used by industry, part processors are required. A part processor is based on the way a specific machine and machine tool controller accepts and understands NC data. There will be a postprocessor for each machine. The postprocessor takes the CL file and converts it into a tape image file. The tape image file can now be used by the CNC operator to machine the required part. To better understand what information is required and how it flows through a CAD/CAM system to generate a CNC tape image file, refer to Fig. 25-1.

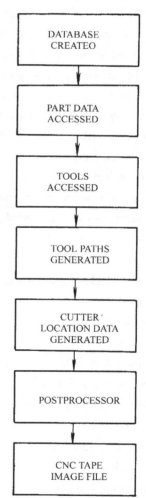

Fig. 25-1 Flow of information required by a CAD/CAM system to generate a CNC tape image file

NC and CAM in the Future

The future of the automated and unattended factory depends more on the manufacturing information programs (software) which will be available and how they will be used than on computer hardware.[2] The factory of the future will depend on NC, which is the key. NC systems will probably improve, but no revolutionary changes are projected in the foreseeable future. The greatest change will come in MCUs which may soon contain the following features:

(1) MCUs may become automatic programming devices, which keep their own program libraries.

(2) MCUs will be able to communicate with other CAM system elements along a local area network (LAN) system.

(3) MCUs will coordinate other slave machines, such as robots, in a manufacturing cell.

(4) MCUs will have diagnostic capabilities enabling them to monitor themselves and the machines they control, to see when maintenance is required, and to report this information.

(5) MCUs will have adaptive control (AC) which can provide feedback information from the cutting tool, workpiece, or machine tool so that the best machining conditions can be maintained.[3] They may sense cutting tool temperature, vibration, a dull cutting tool, motor torque, etc., and automatically change the speeds and feeds to keep tool wear to a minimum while at the same time giving the best metal removal rates.[4]

CNC systems of the future will be less dependent upon the part programmer because the software programs will contain an expert system for generating NC programs. The CAM systems of the future will bring together NC data and other manufacturing information. The NC data combined with group technology (GT), manufacturing process planning (MPP), material requirements planning (MRP), and manufacturing control systems will contain enough information and systems technology for the automated or unattended factory of the future.[5]

Technical Words

perpendicular [ˌpəːpən'dikjulə]	adj. 垂直的
	n. 垂线
profile ['prəufail]	n. 轮廓；仿形
plunge [plʌndʒ]	v. 插入；切入；沉入
graphic ['græfik]	adj. 绘图的
processor ['prəusesə]	n. 处理机，处理器
format ['fɔːmæt, -mɑːt]	n. 形式，格式
image ['imidʒ]	n. 影像
software ['sɔftwɛə]	n. 软件
network ['netwəːk]	n. 网络
diagnostic [ˌdaiəg'nɔstik]	adj. 诊断的
vibration [vai'breiʃən]	n. 振动，摇动
rate [reit]	n. 比率；速度
combine [kəm'bain]	v. 联合，结合
feedback ['fiːdbæk]	n. &v. 反馈

Technical Phrases

adaptive control	自适应控制
cutting tool	刀具
motor torque	电动机转矩
manufacturing cell	制造单元
manufacturing control system	制造控制系统
NC system	数控系统
group technology (GT)	成组技术
MRP (material requirements planning)	物料需求计划

Notes

(1) With a CL file, the information *that was created to generate the graphic display* is converted into coordinate locations to move the cutting tool around the part.

that 引导的定语从句修饰 the information。

(2) The future of the automated and unattended factory **depends more on** the manufacturing information programs (software) *which will be available* and how they will be used **than on** computer hardware.

由 which 引导的定语从句修饰 programs；depend more on... than on... 意为"主要取决于……而不是……"。

(3) MCUs will have adaptive control (AC) *which can provide feedback information from the cutting tool, workpiece, or machine tool* so that the best machining conditions can be maintained.

由 which 引导的定语从句修饰 adaptive control；so that 引出目的状语从句。

(4) They **may sense** cutting tool temperature, vibration, a dull cutting tool, motor torque, etc., and automatically **change** the speeds and feeds to keep tool wear to a minimum while at the same time giving the best metal removal rates.

句中的两个谓语动词为 sense 和 change。

(5) **The NC data** combined with group technology (GT), manufacturing process planning (MPP), material requirements planning (MRP), and manufacturing control systems **will contain enough information and systems technology** for the automated or unattended factory of the future.

句子主干为：The NC data will contain enough information and systems technology.

Exercises

(1) Place a "T" after sentences that are true and an "F" after those that are false.

1) When using CAD/CAM systems to produce CNC data to machine a part, the CNC programmer should create a database at first.

2) The tool library contains a description for each and every tool used during the machining.

3) The CRT screen displays the part drawing and the tool path.

4) The future of the automated and unattended factory depends mainly on how well the part

drawing is created.

5) CNC system of the future will depend more on the part programmer because it can not work without the part program.

(2) Fill in the blanks according to the text with the words and phrases given below. Make change if necessary:

  manufacturing information expert system unattended tool library database

CNC system uses CAD/CAM to produce the required data to machine a part. The programmer can recall the part on the CRT screen after the _____ has been created. A(an) _____ contains a description for all the tools available. The tools required during the machining can be recalled from it. The future of the automated and _____ factory depends more on the _____ program. CNC systems of the future will be less dependent upon the part programmer because a (an) _____ will be used to generate NC programs.

(3) Answer the following questions:

1) What should be done before a part is recalled on the CRT screen?
2) What is the function of a postprocessor on each machine?
3) What is most important for the automated and unattended factory of the future?
4) What is the function of adaptive control?
5) Which capability will enable the MCUs to monitor the machines they control?

(4) Explanation:

  CL file LAN system AC GT MPP

【参考译文】

第 25 课　计算机辅助设计/计算机辅助制造/计算机数字控制

  CAD/CAM 系统能够生成加工零件所需要的 CNC 数据。在准备好刀具清单并为待加工零件编好加工工艺后，CNC 编程员便可开始建立数据库。一旦建好数据库，编程员可在 CRT 显示器上调用这个零件。当显示零件后，编程员可从刀库中选取需要的刀具。刀库中的每一把刀具都有相应的描述，附加一个刀名或标识码。假如，数据库与被选中的刀具是用来进行铣或钻加工的，那么，下一步就是生成刀具轨迹。这种情形下，可以执行的加工分为以下三类：

  钻加工：使 Z 轴与零件表面垂直

  轮廓铣削：铣零件轮廓

  沟槽铣削：使 Z 轴进刀，从零件内部去除材料

  CRT 显示器会显示零件图像与加工该零件的走刀路径。这个图像信息必须再转化为切削位置文件（CL），并由此转换成位置坐标，使刀具围绕零件移动。该信息此时是可读格式的。

  由于工业用机床及其控制器种类繁多，因而需要配备零件处理器。配备零件处理器的基础是专用机床与机床控制器都能够接受并理解 NC 数据。每一台机床都有一个后置处理器，

它能够读取切削位置文件（CL），并将其转换为磁带影像文件。CNC操作员运用磁带影像文件进一步加工需要的零件。为了更好地理解究竟需要什么信息及如何运用这些信息，在CAD/CAM系统上生成CNC磁带影像文件，请参阅图25-1。

NC和CAM的未来

自动化无人工厂的未来主要取决于未来制造软件及其运行情况，而并非计算机的硬件。未来工厂将取决于起关键作用的数控技术。数控系统虽然会有所改进，但在不远的将来还不可能有巨大的变革，改变最大的可能是机床控制单元（MCU）。未来的MCU可能具备以下特点：

（1）MCU可能会成为具有内置程序库的自动编程装置。

（2）MCU通过局域网能与其他CAM组件互通信息。

（3）MCU能协调制造单元中机器人等其他附属机器工作。

（4）MCU将具有自诊断能力，能监控自身与其控制的机床，以便了解何时需要维修，并报告这些信息。

（5）MCU将具有自适应控制（AC）能力，能从切削刀具、工件或机床上提供反馈信息，以保持最佳加工状况；能监测刀具温度、振动、磨损、电动机转矩等，自动改变切削速度与进给量，在保持最佳金属切削效果的同时使刀具磨损最小。

未来的计算机数控系统不会过多地依赖程序员，因为软件程序将包含生成数控程序的专家系统。未来的CAM系统将把NC数据与其他加工信息结合起来。NC数据与成组技术（GT）、制造工艺规划（MPP）、物料需求计划（MRP）、制造控制系统的有机结合，必将给未来自动化无人工厂提供足够的信息与系统技术支持。

Unit 26　Flexible Machining System

Text

A flexible machining system (FMS) is a system of CNC machines, robots, and part transfer vehicles that can take a part from raw stock or casting and perform all necessary machining, part handling, and inspection operations to make a finished part or assembly. It is an entire unmanned, software-based, manufacturing/assembly line. An FMS consists of four major components: the CNC machines, coordinate measuring machines, part handling and assembly robots, and part/tool transfer vehicles (Fig. 26-1).

Fig. 26-1　A small flexible machining system

The main element in an FMS is the CNC machining or turning center. The automatic tool changing capability of these machines allows them to run unattended, given the proper support system. Tool monitoring systems built into CNC machine are used to detect and replace worn tools. The major obstacles in an FMS are not the machining centers but the support systems for the machines, such as part load/unload and part transfer.

Inspection in an FMS is accomplished through the use of coordinate measuring machines. These operate much like CNC machinery in that they are programmed to move to different positions on a workpiece. [1] Instead of using a rotating spindle and a cutting tool, a coordinate measuring machine is equipped with electronic gaging probes which measure features on a workpiece. The results of the gaging are compared to acceptable limits programmed into the machine.

Robots are frequently used in an FMS to load and unload parts from the machines. Since robots are programmed pieces of equipment that lack the ability to make judgments, special workholding fixtures are employed on the transfer vehicles to orient the workpiece so that the robot can handle it correctly. [2] Specially designed machine fixtures and clamping mechanisms are employed to ensure correct placement and clamping of the part on the machine. All part handling must be accomplished in a specific orderly fashion, with coordination of the part transfer vehicle, the robot, and the CNC machines. Future robots will probably employ some type of artificial intelligence which will enable them to make limited judgments as to workpiece orientation and take the necessary corrective actions. [3]

The fourth critical component of an FMS is the part/tool transfer vehicles. These vehicles shuttle workpieces from machine to machine. They also shuttle tool magazines to and from the machinery to maintain an adequate supply of sharp cutting tools at each CNC machine. Transfer vehicles employed in current flexible manufacturing systems are of four major types: automatic guided vehicles (AGV), wire guided vehicles, air cushion vehicles, and hardware guided vehicles.

Automatic guided vehicles rely on onboard sensors and/or a program to determine the path they take. There is no hardware connecting them to the system. An advantage of AGVs is that they can be reprogrammed to take different routes, eliminating the need to run tracks or wires for each route change. The corresponding disadvantage of AGVs is that they are the most difficult of the part delivery vehicles to make function, because of the lack of hardware connection.

A wire guided vehicle uses a wire buried in the floor to define its path. A sensor on the vehicle detects the location of the wire. A major advantage of wire guided vehicles is the ability to use the wire as opposed to an AGV without the need to have a hardware system such as an overhead wire or track on the floor. [4] The disadvantage of wire guided vehicles is the necessity of installing new wire in the floor if a route change is required.

An air cushion vehicle is guided by some external hardware device, such as an overhead wire, but glides on a cushion of air rather than a track system. When using air cushioned vehicles, particular attention to chip removal and control must be built into the FMS. Chips in the path of an air cushion vehicle will stop its progress. These vehicles are generally used for straight paths.

Hardware guided vehicles are the most reliable but least flexible of the transfer vehicles. A track on the floor or an overhead guide rail controls the vehicle path. The advantages of these vehicles are their reliability and the ease of coordinating them with the rest of the system. The major disadvantage is, of course, the need to run new rail or track whenever a vehicle route change or new route is deemed essay. A large FMS may employ several different types of vehicles, depending on the requirements of different parts of the manufacturing line.

Technical Words

vehicle ['viːikl]	n. 搬运装置；传送装置
assembly [ə'sembli]	n. 装配
obstacle ['ɔbstəkl]	n. 障碍
orientation [ˌɔːrien'teiʃən]	n. 定位
artificial [ˌɑːti'fiʃəl]	adj. 人工的
delivery [di'livəri]	n. 传输，传送
sharp [ʃɑːp]	adj. 锋利的，锐利的
sensor ['sensə]	n. 传感器，敏感元件
rail [reil]	n. 铁轨，导轨

Technical Phrases

flexible machining system	柔性制造系统

coordinate measuring machine	坐标测量机
turning center	车削中心
artificial intelligence	人工智能
tool magazine	刀库
automatic guided vehicle	自动导向车
air cushion vehicle	气垫车

Notes

(1) These operate much like CNC machinery in that they are programmed to move to different positions on a workpiece.

These 指文中上句提到的 coordinate measuring machines（坐标测量机）。in that 表示"在……方面"。

(2) Since robots are programmed pieces of equipment *that lack the ability to make judgments*, special workholding fixtures are employed on the transfer vehicles to orient the workpiece so that the robot can handle it correctly.

since 引导原因状语从句，that 引出的定语从句修饰 equipment。

(3) Future robots will probably employ some type of artificial intelligence *which will enable them to **make** limited judgments as to workpiece orientation and **take** the necessary corrective actions*.

which 引导的定语从句修饰 some type of artificial intelligence；该从句中 to make... 和 (to) take... 并列作宾补。

(4) A major advantage of wire guided vehicles is the ability to use the wire as opposed to an AGV without the need to have a hardware system such as an overhead wire or track on the floor.

as opposed to 意为"与……相反"。

Exercises

(1) Place a "T" after sentences that are true and an "F" after those that are false.

1) A flexible machining system can make a finished part from raw stock or casting.

2) The main element in a flexible machining system is part handling robots.

3) Coordinate measuring machines are used for inspection purpose in an FMS.

4) The future robots will probably have the ability to make limited judgments.

5) Automatic guided vehicles can determine the path they take even though there is no hardwire connecting them to the system.

6) It is the sensor on a wire guided vehicle that detects the location of the wire.

(2) Fill in the blanks according to the text with the words and phrases given below. Make changes if necessary.

AGV   unmanned   transfer vehicle   hardware   wire guided   cushion

An FMS is an entire _____, software-based, manufacturing/assembly line. The tool and workpiece _____ are one of its major components. In current flexible manufacturing systems there are four major types of transfer vehicles: _____, _____ vehicles, air _____ vehicles, and _____ guided vehicles. A large FMS may employ several different types of vehicles, depending on the requirements of different parts of the manufacturing line.

（3）Answer the following questions：

1）How many components does an FMS consist of? What are they?

2）In an FMS, how are coordinate measuring machines used to inspect a workpiece?

3）What is employed to help the robot to handle the workpiece correctly?

4）What sort of paths is an air cushion vehicle generally used for?

5）What are the advantages and disadvantages of AGVs?

（4）Explanation：

  FMS AGV

【参考译文】

第26课　柔性制造系统

  柔性制造系统（FMS）是CNC机床、机器人、工件传输装置的集成系统，该系统能从粗坯件或铸件开始，完成所有必要的加工、工件处理与检验，最终形成一个工件或组件成品。它完全是一个基于软件的全自动无人生产线/装配线。柔性制造系统由四个主要部分组成：CNC数控机床、坐标测量机、工件处理与装配机器人、工件/刀具传输装置（见图26-1）。

  柔性制造系统的主要部分是CNC加工中心或车削中心。只要配置适当的支持系统，自动换刀功能使这些机床可在无人监督下运行。CNC机床具有内置刀具监测系统，可监测并更换磨损刀具。柔性制造系统运行的主要障碍不是出在加工中心，而是在机床支持系统，例如工件的装卸与传输等。

  柔性制造系统的检验功能由坐标测量机完成。坐标测量机在程序控制器下沿工件移动，这一点与CNC机床十分相似。不同的是，坐标测量机不用主轴旋转与切削刀具，而是用电子测头在工件上移动来测量工件参数。测量结果要与程序设定的极限值进行比较。

  柔性制造系统常用机器人给机床装卸工件。由于机器人是程控装置，缺乏判断能力，所以传输装置使用专用夹具定位工件，以便机器人能正常运作。使用专门设计的机床夹具与夹紧装置，就是为了确保工件位置正确及在机床上夹紧。工件传输装置、机器人、CNC机床协调工作，所有工件处理必须按特定顺序完成。未来的机器人可能具有某种人工智能，这使它们对工件定位有一定判断能力，且能采取必要的纠错动作。

  柔性制造系统的第四个重要组成部分是工件/刀具传输装置。这些装置使工件在机器之间穿梭移送，也使刀库在机器上来回移动，从而保证每台CNC机床上锋利刀具的正常供给。现代柔性制造系统使用四种主要传输装置：自动导向车（AGV）、线导向车、气垫车和硬件导向车。

  自动导向车依靠安装在车上的传感器和（或）程序来确定其运行路径。没有硬件使其与柔性制造系统相联。自动导向车有一个优点，即人们能程控改变运行路径，无需每次改变路线就要重新布置导轨；相应的缺点是，正因为缺乏硬件联结，自动导向车是最难发挥作用的工件传输装置。

线导向车使用一根埋在地下的导线来确定其路径。车上装有传感器，用于检测导线位置。线导向车的主要优点是，与自动导向车不同，线导向车具有使用导线的能力，但不需要如悬空导线或地上导轨的硬件系统。线导向车的缺点是，如果需要改变行车路线，则需要安装新的地下导线。

气垫车用外部硬件装置（如悬空导线）导向，但它是在气垫上滑行，而不是在导轨上行驶。使用气垫车时一定要注意：柔性制造系统内部必须设置切屑排除与控制功能。气垫车行驶路线上如有切屑，则气垫车无法运行。这种传输装置通常用于直线行车路径。

硬件导向车是传输装置中最可靠但柔性最差的装置。其传输路径由地面导轨或悬空导轨控制。它的优点是可靠、易于和系统其余部分协调。其主要缺点是，每当改变传输路径或想试验一条新路径时，都要架设新的悬空导轨或铺设新的地面导轨。根据生产线上不同部分的需求，一个大型柔性制造系统可能使用多种类型的传输工具。

Unit 27 Employment Opportunities in NC

Text

A number of skilled positions have been created by numerical control. The most common jobs are NC electronic technician, machine operator/setup operator, and part programmer.

Electronic Technician

Numerical control and computer numerical control equipment are electrical systems interfaced to a machine tool. The electronics necessary for a CNC machine to function are complex. The NC electronic technician is a skilled technician who specializes in the maintenance of numerical control equipment. The NC electronic technician must be well trained in digital electronics and possess a knowledge of the cycles and functions of NC machinery. The technician must be able to troubleshoot and correct problems that occur in the electronic circuitry of various NC machines.

NC electronic technicians generally acquire their skills through a two-year junior college program in digital electronics. Additional education in numerical control is often provided by the employer in the form of NC manufacturer's technical school classes and seminars. [1]

Machine Operator/Setup Operator

The machine operator/setup operator is responsible for preparing an NC machine to run a program and for setting up the fixtures, tools, and workpieces. The operator must possess a knowledge of general machine shop practices and techniques, as well as the cycles and functions of an NC machine. The operator is responsible for overriding programmed speeds and feeds if required during machining. The operator also assigns the tool length offsets to the appropriate tool registers and may be called upon to single-step a program through its first cycle. The operator must also be trained in the use of precision measuring instruments as he or she is often responsible for measuring the parts as they are finished.

Machine operators/setup operators acquire their training either by years of running other types of manufacturing equipment and then transferring to an NC operator's position, or through a two-year junior college program. [2] Factory seminars and other coursework may be provided by the employer as required.

Part Programmer

The part programmer is a highly skilled individual responsible for writing the programs that run on numerically controlled equipment. He or she must be trained in general machine shop practice, mathematics, and the use of computers. Based on the part drawing, the programmer selects a machine to machine the part and devises a machining strategy, listing the tools to be used and the coordinates necessary to accomplish the operations. This information is then assembled into a part program written for the particular machine selected.

An NC programmer may acquire training through a two-year junior college, a four-year engineering technology degree program, or by transferring from positions as journeyman machinists or tool and die makers. NC programmers take additional coursework and factory seminars as required by the employer. The educational requirements for a programmer vary with the employer.

Technical Words

interface ['intə(ː)ˌfeis]	v. 连接
maintenance ['meintinəns]	n. 保养，维修
troubleshoot ['trʌblʃuːt]	v. 查找故障；检修故障
seminar ['seminɑː]	n. 研讨会
override [ˌəuvə'raid]	v. 修调
offset ['ɔːfset]	n. 偏移，偏移量
assign [ə'sain]	v. 分配，指定，设定
journeyman ['dʒəːnimən]	n. 熟练工

Technical Phrases

NC electronics technician	数控电子技术员
machine operator	机器操作工
setup operator	安装操作工
part programmer	零件编程员
digital electronics	数字电子学
machine shop	机工车间
electronic circuitry	电子线路
part drawing	零件图
die maker	模具制造工

Notes

(1) Additional education in numerical control is often provided by the employer in the form of NC manufacturer's technical school classes and seminars.

in the form of 意为"以……的形式"; seminar 意为"研讨会"。

(2) Machine operators/setup operators acquire their training either by years of running other types of manufacturing equipment and then transferring to an NC operator's position, or through a two-year junior college program.

by years of 意为"多年的……"。two-year junior college 意为"两年制学院"。

Exercises

(1) Place a "T" after sentences that are true and an "F" after those that are false.

1) The NC electronics technician should specialize in the maintenance of NC equipment.

2) The NC technician can find out what's wrong with the machine and then ask an engineer to repair it.

3) The machine operator must have the ability to use precision measuring instruments because

he or she often uses them to measure the finished parts.

4) The part programmer is responsible for writing programs so he or she can override programmed speeds and feeds if required during machining.

5) The part programmer must have machining practice.

(2) Fill in the blanks according to the text with the words given below. Make changes if necessary:

  precision opportunity coordinate programmer circuitry operator
  fixture maintenance

NC provides a lot of work _____. People who want to be an electronic technician specialize in the _____ of NC equipment and correct problems that occur in the electronic _____ of NC machines. Being a machine _____, he or she must prepare an NC machine to run a program and setup the _____, tools and workpieces. The operator must have the ability to use _____ measuring instruments to measure the finished part. The part _____ is responsible for writing part programs. He or she must list the tools to be used and the _____ necessary, then assembles this information into a part program.

(3) Answer the following questions:

1) Is it necessary for electronic technician to possess a knowledge of the cycle and function of NC machinery?

2) How can NC technician acquire their skills?

3) Why should the NC machine operator be trained to use precision measuring instruments?

4) What knowledge and practice should an NC part programmer have?

5) Who will assign the tool length offsets to the appropriate tool registers?

【参考译文】

第27课　数控职业范畴

  数控技术已经带给人们许多技术岗位，最常见的有 NC 电子技术员、机床操作工/装配工、工件编程员。

电子技术员

  数控（NC）与计算机数控（CNC）设备是与机床接口的电气系统。CNC 机床的运行涉及复杂的电子学知识。NC 电子技术员是专门从事数控设备维护的熟练技师，应受过数字电路的良好培训，并且懂得数控机床功能与整个加工过程。电子技术员要会检修故障、解决各种数控机床电路出现的问题。

  数控电子技术员一般通过两年制学院主修数字电路课程获得岗位技能。此外，还要选学数控知识，通常以雇主提供的数控加工培训班与研讨会的形式学习。

机床操作工/装配工

  机床操作工/装配工负责数控机床的程序准备、安装夹具、装夹刀具及工件。操作工不仅要懂得数控机床功能与整个加工过程，还要有常规机床加工的实践经验与技能。如果加工过程需要，操作工还要负责修调程序中的速度与进给量。操作工也要给相应的刀具存储器指定刀具长度补偿值，并有可能在首次加工循环中单步运行程序。由于要经常测量成品工件，操作工也要受过精密测量仪器使用的培训。

机床操作工/装配工可由两种途径训练所需技能：一是数年操作其他种类加工设备后转到数控操作岗位，二是在两年制学院学习。雇主会视情况举办研讨会与其他进修课程。

工件编程员

工件编程员是技艺精湛的程序员，主要负责编写数控设备运行需要的程序。程序员应具备一般加工经验、数学知识以及计算机应用能力。程序员应根据零件图选择合适的机床加工工件、设计加工方案、列出要用的刀具、算出加工需要的坐标，然后把这些信息汇编到为特选机床编写的程序中去。

数控编程员可经两年制学院培训获得技能，或者攻读四年制工程技术学位获得技能，亦可通过熟练机械师、刀具与模具制造工转岗获得技能。视雇主需要，数控编程员还要进修其他课程、参加工厂研讨会。总之，对编程人员而言，需要的教育因雇主而异。

【Reading Material】

Program Format

Program format, or style, is an important part of CNC machining. Each individual will format his programs differently and, in most cases, a programmer could identify a program written by himself. The point is that a programmer needs to be consistent and efficient, writing code in the way it is listed and in the order it appears in the program. For example:

X, Y, Z is in order of appearance. The machine will read X, Y or Z in any order, but we want to be consistent. Write X first, Y second, Z third.

The first line or block in a program using active G codes should be a tool number and tool change command. This would be a good safety measure.

The second line or block will contain a rapid command (G00), an absolute or incremental command (G90, G91), a work zero for X and Y (G54), a positioning X and Y coordinate, a spindle speed command (S _____), and a spindle ON clockwise command (M03).

The third line or block will contain a "Read tool length compensation" command (G43), a tool length offset number (H01), a Z-axis positioning move (Z0.1), and an optional coolant ON command (M08).

An example program's first three lines will look like this:

T1 M06;
G00 G90 G54 X0 Y0 S2500 M03;
G43 H01 Z0.1 M08;

All the necessary codes for each operation are listed above. This format is a good practice and will separate your style from other programmers.

QUESTION:

If G00, G90, and G54 are defaults, why do we list them in the second line of a program and for each different tool?

ANSWER:

G00, G90, and G54 are listed for an operator/setup person's aid. They show that the machine will rapid-position, that the machine is in the absolute coordinate mode, and most important, the work zero coordinates. The work zero is always different between setups, and multiple work zeros are very common.

QUESTION:

Can we combine the second and third lines, excluding the M08 code? If so, why do we write the lines separately?

ANSWER:

Yes. The four G codes G00, G90, G54, and G43 all belong to different groups. Remember, no two G codes of the same group can be listed on the same line. However, the main reason for using two lines is SAFETY.

Remember, only one line of information can be executed at a time. The X and Y coordinates will position first, then the tool length and the Z coordinate will execute. If combined, all three axes will move simultaneously, and any interfering clamps or fixtures can be struck and/or destroyed. When combining X, Y and Z in positioning, chances of crashing the machine are greater.

Unit 28 An Introduction to Industrial Robots

Text

The term *robot* and *robotics* were created well before robots became the reality. The term robots first entered the English language with the translation of Czechoslovakian playwright Karel Capek's play *R. U. R.* (*Rossum's Universal Robots*), in 1923. The word *robot* comes from the Czech robot that means slave or worker. In 1942 another author Isaac Asimov (who wrote many short stories about robots), created the term *robotics* when he established his Three Laws of Robotics. Robots, he reasoned would have special circuitry to make the robot always conform to these basic laws:

(1) A robot must not harm a human being, nor through inaction allow one to come to harm.

(2) A robot must always obey human beings, unless that is in conflict with the first law.

(3) A robot must protect itself from harm, unless that is in conflict with the first or second law.

When these stories were written they were pure science fiction. Today with robots a reality, analyzing these laws of robotics provides a worthwhile concept for roboticists to consider when designing robot control systems.[1]

Robot Definition

A robot is a re-programmable, multi-functional manipulator designed to move materials, parts, tools, special devices through variable programmed motions for the performance of a variety of tasks.

This definition is universally accepted. The main point is that industrial robots are re-programmable, and are capable of different types of path movements.

The Development of Robots

With the development of numerically controlled machines the concept of imitating a human arm to manipulate parts became a natural progression.

Contrary to common belief, robotics was not developed recently. In fact, the first robots were created in the early 1960's in America. Unimation produced a robot arm in 1961 with the control unit sequence set by the operator. But due to the experimental nature of the work a low profile was kept mainly to avoid adverse public reaction to the project.[2] In 1974 Cincinnati Millicron was the first mini-computer controlled robot. However, in the same year the IRB6 robot was introduced by the Swedish company ASEA. This robot has been marketed all over the world and is still in production today (1991) with the only major changes being the upgrading of the control cabinet electronics and software.[3] So while America may be credited for establishing the technology for robotics, countries like Japan and Sweden have utilized it to a greater extent in industrial applications.

Many reasons can be attributed to the increase in awareness and use of industrial robots in recent time; some being the ever increasing cost of unskilled labor, better design and performance, the rapid advances in computer and sensory technology, the desire to take employees away from mundane, repetitive or potentially dangerous work environments.

Technical Words

robotics [rəu'bɔtiks]	n.	机器人学；机器人技术
roboticist [rəu'bɔtisist]	n.	机器人专家
imitate ['imiteit]	v.	模仿，仿制
manipulator [mə'nipjuleitə]	n.	操纵器；操作机
manipulate [mə'nipjuleit]	v.	操纵，操作
upgrade ['ʌpgreid]	v. & n.	升级

Notes

(1) Today with robots a reality, analyzing these laws of robotics provides a worthwhile concept for roboticists to consider when designing robot control systems.

句子主干为：Analyzing these laws of robotics provides a worthwhile concept.

(2) But due to the experimental nature of the work **a low profile was kept** mainly to avoid adverse public reaction to the project.

profile 意为"仿形"。due to the experimental nature of the work 意为"鉴于这项工作尚属试验"。句子主干结构为：A low profile was kept.

(3) This robot has been marketed all over the world and is still in production today (1991) with the only major changes being the upgrading of the control cabinet electronics and software.

market 意为"销售"；upgrade 意为"升级"；all over the world 意为"全世界"。

Exercises

(1) Place a "T" after sentences that are true and an "F" after those that are false.

1) The term *robot* was not created well before robots became true.

2) Since a robot must obey human beings, people can design robot control systems for any purpose.

3) According to the Three Basic Laws of Robotcs, a robot should be harmless to a human being.

4) One reason that robots are increasingly used is that computer technology is advancing rapidly.

(2) Fill in the blanks according to the text with the words given below. Make changes if necessary.

  harm program reality dangerous obey robot

When the term _____ entered the English language, robots had not become the _____. Authors described robot in their stories. They thought a robot should always _____ human beings, and must not _____ a human being. Industrial robots are designed to perform a variety of tasks after being _____. Nowadays, robots are widely used in industry because they can take employees away from repetitive or _____ work environments.

(3) Answer the following questions:

1) What is the meaning of the word *robot* when it first appeared in authors stories?
2) Did robot become the reality as soon as the term was created?
3) When designing a robot, what laws should be considered by the designer?
4) When was the robot produced?
5) Why is the use of robot increasing?

【参考译文】

第28课 工业机器人入门

早在机器人变为现实之前，机器人与机器人学这两个术语就已经提出来了。1923年，随着捷克剧作家卡雷尔·查培克的剧本《R. U. R》(《罗苏姆的通用机器人》) 英文译本的问世，机器人这一术语开始进入英语。机器人 robot 一词源于捷克语，该词意指奴隶或劳工。1942年，另一位作家艾萨克·埃思穆乌（他曾撰写过许多有关机器人的短篇小说）在创立其机器人学三个法则时就提出了机器人学这个术语。他曾推断，机器人应有特殊电路，使其始终遵循下述三条基本法则：

(1) 机器人不能伤害人类，也不能通过不执行指令而使人类受到伤害。
(2) 在不违背第一条法则的前提下，机器人必须始终遵从人类意志。
(3) 在不违背第一、二条法则的前提下，机器人必须保护自身不受伤害。

当时撰写的这些故事纯属科学幻想。今天，随着机器人变成现实，分析这些机器人法则，从中获得很有价值的理念，可供机器人专家设计机器人控制系统时参考。

机器人的定义

机器人是一种可重复编程的多功能操纵器，其设计用途是输送物料、工件、刀具及一些特殊装置，通过各种程控运动来完成多种不同任务。

以上定义被普遍认可，其要点是：工业机器人可以重复编程，且能沿多种不同轨迹运动。

机器人的发展史

随着数控机床的发展，模仿人类手臂操纵工件的想法便自然地提出来了。

与常规观点相反，机器人学并非最近发展起来的。事实上，早在20世纪60年代初期，美国人便制造出第一批机器人。Unimation（万能自动化）公司于1961年就生产出机械手臂，其控制装置的时序是由操作者预设的。然而，鉴于这项工作尚属试验，为了避免公众对该项目的抵触情绪，当时的仿形程度较低。1974年，辛辛那提 Millicron 机器人成为首例以小型计算机控制的机器人。然而，就在同一年里，瑞典 ASEA 公司推出了它的 IRB6 机器人。这种机器人一直在全球畅销，现在（1991年）还在生产，唯一的重大改进是控制柜电子装置与软件的升级。所以，当人们以为美国正在建立机器人技术的时候，像日本与瑞典这样一些国家，机器人在工业中的应用已经达到很高水平。

近来有许多原因促使人们越来越意识到应用工业机器人的重要性，其中有些原因是日益增加的花费所致：例如培训新工人的费用、改进设计与产品性能的花费、计算机与传感器技术飞速发展导致的费用，以及雇员们希望摆脱平淡、重复、有潜在危险工作环境的花费等。

Unit 29　Computer Integrated Manufacturing System

Text

Definition of CIM

Computer integrated manufacturing (CIM) is the term used to describe the most modern approach to manufacturing. Although CIM encompasses many of the other advanced manufacturing technologies such as computer numerical control (CNC), CAD/CAM, robotics and just-in-time delivery (JIT), it is more than a new technology or a new concept. [1] Computer integrated manufacturing is actually an entire new approach to manufacturing or a new way of doing business.

To understand CIM, it is necessary to begin with a comparison of modern and traditional manufacturing. Modern manufacturing encompasses all of the activities and processes necessary to convert raw materials into finished products, deliver them to the market, and support them in the fields. [2] These activities include the following:

(1) Identifying a need for a product.

(2) Designing a product to meet the needs.

(3) Obtaining the raw materials needed to produce the product.

(4) Applying appropriate processes to transform the raw materials into finished product.

(5) Transporting product to the market.

(6) Maintaining the product to ensure a proper performance in the field.

This broad, modern view of manufacturing can be compared with the more limited traditional view that focuses almost entirely on the conversion processes. [3] The old approach separates such critical pre-conversion elements as market analysis research, development, and design for manufacturing, as well as such after-conversion elements as product delivery and product maintenance. In other words, in the old approach to manufacturing, only those processes that take place on the shop floor are considered manufacturing. This traditional approach of separating the overall concept into numerous stand-alone specialized elements was not fundamentally changed with the advent of automation. While the separate elements themselves became automated (i.e. computer-aided drafting and design (CAD) in design and CNC in machining), they remained separate. Automation alone did not result in the integration of these islands of automation.

With CIM, not only are the various elements automated, but also the islands of automation are all linked together or integrated. [4] Integration means that a system can provide complete and instantaneous sharing of information. In modern manufacturing, integration is accomplished by computers. With this background, CIM can now be defined as the total integration of all manufacturing elements through the use of the computers.

Historical Development of CIM

It has taken many years for CIM to develop as a concept, but integrated manufacturing is not really new. In fact, integration is where manufacturing actually began. Manufacturing has evolved through the following four distinct stages:

- **Manual Manufacturing**

Manual manufacturing using simple hand tools was actually integrated manufacturing. All information needed to design, produce, and deliver a product was readily available because it resided in the mind of the person who performed all of the necessary tasks. [5] The tool of integration in the earliest years of manufacturing was the human mind of the craftsman who designed, produced, and delivered the product.

- **Mechanization/Specialization**

With the advent of the industrial revolution, manufacturing processes became both specialized and mechanized. Instead of one person designing, producing, and delivering a product, workers and machines performed specialized tasks within each of these broad areas. Communication among these separate entities was achieved by using drawings, specifications, job orders, process plans, and a variety of other communication aids. [6]

- **Automation**

Automation improved the performance and enhanced the capabilities of both people and machines within specialized manufacturing components. For example, CADD enhanced the capability of designers and drafters; CNC enhanced the capability of machinists; and computer-assisted process planning (CAPP) enhanced the capabilities of industrial planners.

- **Integration**

With the advent of the computer age, manufacturing has developed full circle. It began as a totally integrated concept and, with CIM, has once again become one. However, there are major differences in the manufacturing integration of today and that of the manual era of the past. First, the instrument of integration in the manual era was the human mind. The instrument of integration in modern manufacturing is the computer. Second, processes in the modern manufacturing setting are still specialized and automated.

Technical Words

approach [əˈprəutʃ]	n. 方法，方式
robotics [rəuˈbɔtiks]	n. 机器人；机器人学
integration [ˌintəˈgreiʃən]	n. 集成；整合；积分
mechanization [ˌmekənaiˈzeiʃən]	n. 机械化
specialization [ˌspeʃəlaiˈzeiʃən]	n. 专门化
specification [ˌspesifiˈkeiʃən]	n. 规格，规范
drafter [ˈdrɑːftə]	n. 草图设计员
machinist [məˈʃiːnist]	n. 机械师；机工

Technical Phrases

just-in-time delivery (JIT)	准时生产

finished product	成品
shop floor	生产现场
job order	工作任务单
process plan	工艺规划
Computer-assisted Process Planning (CAPP)	计算机辅助工艺规划

Notes

(1) Although CIM encompasses many of the other advanced manufacturing technologies such as computer numerical control (CNC), CAD/CAM, robotics and just-in-time delivery (JIT), it is more than a new technology or a new concept.

在 it is more than a new technology or a new concept 中，代词 it 指的是 CIM；more than 意为"远不止于……"。

(2) Modern manufacturing encompasses all of the activities and processes necessary to convert raw materials into finished products, deliver them to the market, and support them in the fields.

句中的 to convert raw materials into finished products 意思是"将原材料加工成成品"，deliver them to the market 意为"把产品推向市场"，support them in the fields 指的是"保障产品能在实际中正常使用"。

(3) This broad, modern view of manufacturing can be compared with the more limited traditional view *that focuses almost entirely on the conversion processes*.

句中的 that focuses almost entirely on the conversion processes 为定语从句，修饰 traditional view；conversion processes 是指将原材料加工成成品的"加工过程"。

(4) With CIM, not only are the various elements automated, but also the islands of automation are all linked together or integrated.

句中的 not only are the various elements automated 是倒装强调句。the various elements 是主语，谓语动词 are 倒装在主语之前。

(5) All information needed to design, produce, and deliver a product was readily available because it resided in the mind of the person *who performed all of the necessary tasks*.

句中的 who performed all of the necessary tasks 是定语从句，修饰 the person；定语从句的意思是"完成所有任务的人"，根据上下文可不译出。

(6) Communication among these separate entities was achieved by using drawings, specifications, job orders, process plans, and a variety of other communication aids.

特定任务之间的沟通是通过图样、技术规范、任务单、工艺规划以及其他手段实现的。

Exercises

(1) Place a "T" after sentences that are true and an "F" after those that are false.

1) Computer integrated manufacturing is just a new technology or a new concept in modern society.

2) In modern approach to manufacturing, only those processes that take place on the shop floor

are considered manufacturing.

3) With CIM, not only are the various elements automated, but also the islands of automation are all linked together or integrated.

4) The instrument of integration in modern manufacturing is the human mind.

(2) Fill in the blanks according to the text with the words given below. Make changes if necessary.

technology   integration   approach   computer   mind

CIM is the term used to describe the most modern _____ to manufacturing. CIM encompasses many of the other advanced manufacturing _____.

There are two major differences in the manufacturing integration of today and that of the manual era of the past. First, the instrument of _____ in the manual era was the human _____, while the instrument in modern manufacturing is the _____. Second, processes in the modern manufacturing setting are still specialized and automated.

(3) Answer the following questions:

1) What is the meaning of CIM?

2) How many activities does the modern manufacturing include? What are they?

3) What is the main difference between modern manufacturing and traditional manufacturing?

4) How many distinct stages has manufacturing evolved through?

【参考译文】

第29课　计算机集成制造系统

计算机集成制造的定义

CIM 是一个术语，用来描述"计算机集成制造"这一最先进的制造方法。尽管 CIM 涉及计算机数字控制、计算机辅助设计与制造、机器人技术、准时生产等其他先进制造技术，但它的意义远超出了新技术、新概念的范畴。CIM 实际上已经成为一种全新的制造方法或贸易方式。

为进一步了解 CIM，有必要从现代制造业与传统制造业的对比讲起。现代制造业涉及产品加工、市场营销、现场服务的所有活动与过程，这些活动包括：

（1）确认对某产品的需要。

（2）设计满足要求的产品。

（3）获得生产工件的原料。

（4）使用正确工艺把原料加工为成品。

（5）把产品运送到市场进行销售。

（6）保养产品，确保现场使用性能。

与广义的现代制造业相比，传统的制造业几乎完全集中在产品的加工过程上，而把前期的市场调研与开发、产品设计这些关键环节以及后期的货物运送、产品保养环节完全分割开来。换言之，传统制造业仅仅把车间里的加工环节视为制造。这种把完整的制造体系分割为若干独立的专项环节的传统制造业并没有随着自动化的出现而发生根本性改变。虽然这些分割的环节各自已经自动化（如计算机辅助绘图与设计使得设计过程实现自动化，计算机辅助数控加工使得加工过程实现自动化），但这些环节依然还是分割开来的。也就是说，自动化本身并没有给这些"分块自动化"的环节带来集成化。

CIM 的出现，不仅使得制造过程的各个环节自身实现自动化，而且所有分块自动化的环节都相互联系起来，也就是实现了集成化。所谓集成化，就是指一个系统可以提供完整的即时信息共享。在现代制造业中，集成化是由计算机来实现的。在此背景下，我们可以这样定义 CIM，即运用计算机把所有制造环节完整地集成起来。

计算机集成制造的历史变迁

计算机集成制造的概念经历了多年发展，但集成制造这一概念并不新鲜。事实上，制造业就是从集成化开始的。制造业经历了以下四个不同的发展阶段：

- 手工制造阶段

使用简单手工工具的手工制造业实际上就是一种集成制造，因为有关产品设计、生产、输送所需要的全部信息就在完成所有加工任务的那个工匠心中。早期制造业的集成工具就是设计、生产、输送产品的工匠的智慧。

- 机械化/专业化阶段

随着工业革命的出现，制造过程实现了专业化和机械化。产品的设计、制造、输送不再由一位匠人完成，而是由一些工人和机器完成每个阶段特定任务来实现。这些特定任务之间的沟通是通过图样、技术规范、任务单、工艺规划以及其他手段实现的。

- 自动化阶段

自动化改进了产品性能，也提高了人与机器在各项专门加工环节中的生产能力。例如，使用计算机辅助绘图与设计强化了设计师和绘图员的工作效率，使用计算机数字控制技术增强了机械工人的加工能力，使用计算机辅助工艺规划增强了工业规划师的设计能力。

- 集成化阶段

随着计算机时代的到来，制造业发展正好经历了一个循环。虽然制造业是从整体集成化这一概念开始的，但只有 CIM 的使用才使得集成化变成一个整体。但是，当今的制造业集成化与过去手工时代制造业集成化存在着较大差异。首先，手工制造业的集成工具是人的思维，而当今制造业的集成工具则是计算机；其次，现代制造环境下的制造过程仍然是专业化和自动化的。

【Reading Material】

Problems Associated with CIM

1. Technical Problems of CIM

As each island of automation began to evolve, specialized hardware and software for that island were developed by a variety of producers. This led to the same type of problem that has been experienced in the automotive industry. One problem in maintaining and repairing automobiles has always been the incompatibility of spare parts among various makes and models. Incompatibility summarizes in a word the principal technical problem inhibiting the development of CIM. Consider the following example. Supplier A produces hardware and software for automating the design process. Supplier B produces hardware and software for automating such manufacturing processes as machining, assembly, packaging, and materials handling. Supplier C produces hardware and software for automating processes associated with market research. This means a manufacturing firm may have three automated components, but on systems produced by three different suppliers. Consequently, the three systems are not compatible. They are not able to communicate among themselves. Therefore, there can be no integration of the design, production, and market research, processes.

An effort known as manufacturing automation protocol (MAP) is beginning to solve the incompatibility of hardware and software produced by different suppliers. As MAP continues to evolve, the incompatibility problem will eventually be solved and full integration will be possible among all elements of a manufacturing plant.

2. Cultural Problems of CIM

Computer integrated manufacturing is not just new manufacturing technology; it is a whole new approach to manufacturing, a new way of doing business. As a result, it involves significant changes, for people who were educated and are experienced in the old ways. As a result, many people reject the new approach represented by CIM for a variety of reasons. Some simply fear the change that it will bring in their working lives. Others feel it will altogether eliminate their positions, leaving them functionally obsolete. In any case, the cultural problems associated with CIM will be more difficult to solve than the technical problems.

3. Business-related Problems of CIM

Closely tied the cultural problems are the business problems associated with CIM. Prominent among these is the accounting problem. Traditional accounting practices do not work with CIM. There is no way to justify CIM based on traditional accounting practices. Traditional accounting practices base cost-effectiveness studies on direct labor savings whenever a new approach or new technology is proposed. However, the savings that result from CIM are more closely tied to indirect and intangible factors, which are more difficult to quantify. Consequently, it can be difficult to convince traditional business people, who are used to relying on traditional accounting practices, to see that CIM is an approach worth the investment.

Unit 30 Manufacturing Technology Facing the 21st Century

Text

If a manufacturing company is going to be successful in the 21st century, being good at just "the technology" is not enough to survive. A company must be alert to change; it must offer its customers the most innovative product at the best price and the best all-around service. [1]

Agile Manufacturing

Agility is dynamic, context-specific, aggressively change-embracing, and growth-oriented. It is not about improving efficiency, cutting costs, or battening down the business hatches to ride out fearsome competitive "storms". [2] It is a comprehensive response to the business challenges of profiting from rapidly changing, continually fragmenting, global markets for high quality, high performance, customer configured goods and services.

Offering individualized products—not a bewildering list of options and models but a choice of ordering a product configured by the vendor to the particular requirements of individual customers—is the feature of agile competition. [3]

The ability of companies to fragment markets, to build to order in arbitrary lot sizes, to widen their product ranges and change models frequently, to customize mass-market products, and to market information has undermined the competitiveness of the mass-production system. The acquisition of these abilities is an expression of industry "backing into" agility, unwittingly. [4]

As a comprehensive system, agility defines a new paradigm for doing business. It reflects a new mind-set about making, selling, and buying, openness to new forms of commercial relationships and new measures for assessing the performance of companies and people.

Rapid Prototyping and Manufacturing

To substantially shorten the time for developing patterns, moulds, and prototypes, some manufacturing enterprises have started to use rapid prototyping (RP) methods for complex patterns making and component prototyping. Over the past few years, a variety of new rapid manufacturing technologies, generally called Rapid Prototyping and Manufacturing (RP&M), have emerged; the technologies developed include Stereolithography (SL), Selective Laser Sintering (SLS), Fused Deposition Modeling (FDM), Laminated Object Manufacturing (LOM), and Three Dimensional Printing (3-DPrinting). These technologies are capable of directly generating physical objects from CAD databases. They have a common important feature: the prototype part is produced by adding material rather than removing materials, that is, a part is first modeled by a geometric modeler such as a solid modeler. And then is mathematically sectioned (sliced) into a series of parallel cross-section pieces. [5] For each piece, the curing or binding paths are generated. These curing or binding paths are directly used to instruct the machine for producing the part by solidifying or binding a line of material. After a layer is built, a new layer is built on the previous one in the same way. Thus, the model is built layer by layer from the bottom to top. In summary, the rapid prototyping activities consist of two parts: data preparation and model production.

Powerful new lasers may also open doors to direct manufacturing. Such laser systems are being explored at national laboratories such as Sandia and Los Alamos, as well as at the University of Michigan, Penn State, and elsewhere. They may soon be available commercially. In the Sandia system, a 1,000-watt neodymium YAG (yttrium-aluminum-gallium) laser melts powdered materials such as stainless and tool steels, magnetic alloys, nickel-based superalloys, titanium, and tungsten in layers to produce the final part. The process is slow: three hours to make a one-cubic-inch object. But the part is just as metallically dense as one made by conventional means. Sandia vice president Robert J. Eagan says the lab's researchers hope to see the process used to make replacement parts for the military's stored nuclear weapons. Commercial interest is high too. Ten companies, including Allied Signal and Lockheed Martin, are participating in the program. Another 20 companies support research at Penn State, where the goal is to make big objects, such as tank turrets and portions of airplanes, as a single part.

Environmentally Conscious Design and Manufacturing

In the United States, the municipal solid waste (MSW) generated by households and industrial establishments is about 4 pounds per person each day. According to a current report, the United States has lost more than 70% of its landfill sites in the past 10 years.[6] The report also infers that landfills in many states are reaching their permitted capacities. Facing this environmental problem, both the government and industrial companies are making more strict regulations to promote environmentally friendly products and technology. For example, the governments of Germany and the US require that manufacturers take responsibility for the disposal of their products. The Green Plan of Canada was proposed in 1990 to reduce the stabilization of CO_2 and other greenhouse emissions by the year 2000. Some governments have set up official eco-labeling schemes, intended to inform customers of environmentally friendly products.[7] All of these regulations intend to minimize the environmental impact of products.

Products affect the environment at many points in their lifecycles. These environmental effects result from the interrelated decisions made at various stages of a product's life. Once a product moves from the drawing board into the production line, its environmental attributes are largely fixed. Therefore, it is necessary to support the design function with tools and methodologies that enable an assessment of the environmental consequences (such as emissions, exposure, and effects) in each phase.[8] Environmentally conscious design and manufacturing (ECD&M) is a new view of manufacturing that includes the social and technological aspects of the design, synthesis, processing, and use of products in continuous or discrete manufacturing industries. The benefits of ECD&M include safer and cleaner factories, worker protection, reduced future costs for disposal, reduced environmental and health risks, improved product quality at lower cost, better public image, and higher productivity. Environmentally conscious technologies and design practices will also allow manufacturers to minimize waste and to turn waste into a profitable product.

Technical Words

dynamic [dai'næmik]	adj. 动态的；动力学的
embrace [im'breis]	v. 包含；欣然接受
acquisition [ˌækwi'ziʃən]	n. 获得，取得
paradigm ['pærədaim]	n. 范例，模范
agility [ə'dʒiləti]	n. 敏捷；灵活
response [ri'spɔns]	n. 反应，响应
pattern ['pætən]	n. 图案；样式；样品
mould [məuld]	n. 模子
prototype ['prəutətaip]	n. 原型；标准
component [kəm'pəunənt]	n. 零件；成分
stereolithography	n. 立体平版印刷
database ['deitəˌbeis]	n. 数据库
neodymium [ˌniːəu'dimiəm]	n. 钕
yttrium ['itriəm]	n. 钇
aluminum [ə'luminəm]	n. 铝
gallium ['gæliəm]	n. 镓（金属元素）
titanium [tai'teniəm]	n. 钛（金属元素）
tungsten ['tʌŋstən]	n. 钨
superalloy [ˌsupə'ælɔi]	n. 超合金
stabilization [ˌsteiblai'zeiʃən]	n. 稳定
methodology [ˌmeθə'dɔlədʒi]	n. 方法论
conscious ['kɔnʃəs]	adj. 有意识的，觉察的

Technical Phrases

Agile Manufacturing	敏捷制造
Rapid Prototyping and Manufacturing	快速成型与制造
rapid prototyping (RP)	快速成型技术
Selective Laser Sintering (SLS)	选择性激光烧结
Fused Deposition Modeling (FDM)	熔融堆积成型法
Laminated Object Manufacturing (LOM)	分层实体技术
Three Dimensional Printing (3-DPrinting)	三维打印技术
solid modeler	实体建模，实体模型器
YAG (yttrium-aluminum-gallium) laser	钇铝石榴石激光
tool steel	工具钢
magnetic alloy	磁性合金

nickel-based superalloy	镍基超级合金
Environmentally Conscious Design and Manufacturing	绿色设计与制造
municipal solid waste (MSW)	城市固体废物
greenhouse emission	温室气体

Notes

(1) A company must be alert to change; it must offer its customers the most innovative product at the best price and the best all-around service.

alert to change 意为"对市场变化具有敏锐性", all-around service 意为"周到的服务", at the best price 意为"以最好的价格"。

(2) Agility is dynamic, context-specific, aggressively change-embracing, and growth-oriented. It is not about improving efficiency, cutting costs, or battening down the business hatches to ride out fearsome competitive "storms".

batten down the hatches 原意是风暴到来之前"将船舱封闭起来"; ride out 意为"安然躲过"。

(3) Offering individualized products—not a bewildering list of options and models but a choice of ordering a product configured by the vendor to the particular requirements of individual customers—is the feature of agile competition.

此句比较长,但 Offering individualized products is the feature of agile competition 为句子的主干成分; individualized products 意为"个性化产品"。

(4) The acquisition of these abilities is an expression of industry "backing into" agility, unwittingly.

back into 意为"支持"。

(5) And then is mathematically sectioned (sliced) into a series of parallel cross-section pieces.

mathematically 意为"通过数学算法"。

(6) According to a current report, the United States has lost more than 70% of its landfill sites in the past 10 years.

landfill sites 意为"垃圾填埋场"。

(7) Some governments have set up official eco-labeling schemes, intended to inform customers of environmentally friendly products.

eco-labeling scheme 意为"环保标签方案"; environmentally friendly product 意为"环保产品"。

(8) Therefore, it is necessary to support the design function with tools and methodologies that enable an assessment of the environmental consequences (such as emissions, exposure, and effects) in each phase.

environmental consequence 意为"环保影响"; assessment 指的是"环境评价"。

Exercises

(1) Place a "T" after sentences that are true and an "F" after those that are false.

1) A successful manufacturing company in the 21st century must be only good at "the technology".

2) Agility is static, context-specific, aggressively change-embracing, and growth-oriented.

3) The prototype part is produced by adding material rather than removing materials.

4) Environmentally conscious technologies and design practices will allow manufacturers to minimize waste and to turn waste into a profitable product.

(2) Fill in the blanks according to the text with the words given below. Make changes if necessary.

conscious   prototyping   agile   innovative   alert

A manufacturing company in the 21st century must be _____ to change; it must offer its customers the most _____ product at the best price and the best all-around service. In order to share the market, it must have the ability of _____ manufacturing. To substantially shorten the producing time, some manufacturing enterprises have started to use rapid _____ methods. Meanwhile, they also should consider the influence on environment. Environmentally _____ technologies and design practices will also allow manufacturers to minimize waste and to turn waste into a profitable product.

(3) Answer the following questions:

1) What is the definition of agile manufacturing?

2) How many technologies does the rapid prototyping and manufacturing include? What are they?

3) Why must a manufacturing company in the 21st century consider the influence on the environment?

【参考译文】

第30课　面向21世纪的制造技术

如果制造公司想在21世纪走向成功，仅仅依靠技术优势是难以生存的。要生存，公司就必须要有对市场变化的敏锐性，必须以最好的价格、最周到的服务为公司客户提供最有创意的产品。

敏捷制造

敏捷是企业在特殊背景下着眼未来发展、主动适应变革的一种动态特质，而不仅仅是提高效率、降低成本，或用回避方式躲过市场的激烈竞争风暴。敏捷也是企业应对各种经营挑战的综合对策，以求在瞬息万变、持续分化的全球市场中，提供高质量、高绩效、符合客户需求的商品与服务，从而获取利益。

提供个性化产品——由供应商订制产品以满足个体客户的特定需求，而不是提供眼花缭乱的选项清单——这就是敏捷竞争的特点。

公司在分拆市场份额、批量订购商品、扩展产品系列、及时更新款式、订制畅销商品、掌握市场信息等方面的能力，削弱了大批量生产系统的竞争力。公司对这些能力的获得无意间表达了工业界对敏捷的支持。

敏捷制造是一个新的综合经营模式，它折射出关于制造、销售、购买的一种新理念，对新型商业关系的一种开放态度，人与公司业绩评价的一种新的衡量标准。

快速成型与制造

为大幅缩短样品、模具、原型制作时间，一些制造企业开始使用快速原型法制作复杂样品和部件原型。过去几年已出现一些新的快速制造技术（通常叫做快速成型与制造），其中包括立体平版印刷技术、选择性激光烧结技术、熔融堆积成型技术、分层实体技术以及立体印刷技术（三维打印技术）。这些技术能够由CAD数据库直接生成实体。这些技术都有一个重要的特点：原型工件是用添加材料而不是去除材料的方法形成，也就是说首先用诸如实体建模器的几何造型软件构建部件模型，然后用数学方法将模型切成一系列平行的横截面切片，每个切片产生一些固化或粘结迹线，这些迹线可用来直接引导机床通过固化或粘结材料上的基准线生成零件；建起一层后，用相同方法在前一层上建立新的一层；模型就是这样一层一层自下而上地建立起来的。总之，快速原型法由两部分组成：数据准备和建立模型。

新的激光技术也为直接加工打开了通道。这些激光系统正在被诸如桑迪亚国家实验室和洛斯阿拉莫斯国家实验室、密歇根大学、宾夕法尼亚州立大学等机构进行探索研究，也许很快可付诸应用。桑迪亚国家实验室研究的1千瓦钕钇铝石榴石激光设备，可将诸如不锈钢及工具钢、磁性合金、镍基超级合金、钛、钨的粉状材料熔化成薄片，以便生产成品工件。设备加工过程比较慢，三小时才能加工出1立方英寸的实体。但工件的金属密度与传统制造的一样好。该实验室副主任Robert J. Eagan说，实验室研究人员希望将这项制造工艺用在制造军用核武器备件上。此外，商界对此也很感兴趣。已有10家公司（包括Allied Signal和Lockheed Martin）加入了项目研发，另有20家公司支持宾州的研究项目，该项目的目标是制造诸如坦克炮塔、飞机部件等大型工件。

绿色设计与制造

美国家庭和厂商产生的城市固体废弃物大约每人每天4磅左右。根据近期报告，过去10年间，美国已用掉70%多的垃圾填埋场。该报告推测，美国许多州的垃圾填埋场已达到设计容量。面对这样的环境问题，政府和公司都在制定严格法规，促使环保型产品和技术的开发与运用。比如，美国和德国政府要求制造商对它们产品的处置要负起责任。1990年加拿大绿色环保规划提出，到2000年降低二氧化碳及其他温室气体排放量。一些国家的政府甚至制定了官方环保标签方案，目的是提示客户使用绿色产品。所有这一切措施，试图将产品对环境的影响降到最低限度。

产品在其生命周期的许多阶段都会影响环境，这些影响源于产品生命过程中各阶段所做的相关决策。产品一旦从绘图板走进生产线，其环境属性基本就确定了。因此，必须使用合适的刀具与加工方法保证产品设计功能，使其在每个阶段通过环境评价（如气体排泄、环境风险等）。绿色设计与制造可以说是工业制造的一个新视角，它从社会层面与技术层面涵盖了连续或离散制造工业中产品的设计、综合、加工及应用。绿色设计与制造的优点是：生产环境安全清洁、职工劳动保护良好、降低处理成本、降低环境及健康风险、提升产品质量、降低产品成本、良好的公众形象以及较高的生产效率。绿色技术与设计还可以使制造商减少废料并把废料变成可盈利产品。

【Reading Material】

Nanomaterial and Nanotechnology

Nanomaterials and nanotechnology have become a magic word in modern society. Nanomaterials represent today's cutting edge in the development of novel advanced materials which promise tailor-made functionality and unheard applications in all key technologies. So nanomaterials are considered as a great potential in the 21th century because of their special properties in many fields such as optics, electronics, magnetics, mechanics, and chemistry.

Research on nanomaterials has been stimulated by their technological applications. The first technological uses of these materials were as catalysts and pigments. The large surface area to volume ratio increases the chemical activity. Because of this increased activity, there are significant cost advantages in fabricating catalysts from nanomaterials.

In addition to technology, nanomaterials are also interesting systems for basic scientific investigations. New techniques have been developed recently that have permitted researchers to produce larger quantities of other nanomaterials and to better characterize these materials.

Nanotechnology is a term that entered into the general vocabulary only in the late 1970's, mainly to describe the metrology associated with the development of X-ray, optical and other very precise components. We defined nanotechnology as the technology where dimensions and tolerances in the range 0.1nm-100nm (from the size of the atom to the wavelength of light) play a critical role.

This definition is too all-embracing to be of practical value because it could include, for example, topics as diverse as X-ray crystallography, atomic physics and indeed the whole of chemistry. So the field covered by nanotechnology is later narrowed down to manipulation and machining within the defined dimensional range (from 0.1nm to 100nm) by technological means, as opposed to those used by the craftsman, and thus excludes, for example, traditional forms of glass polishing. The technology relating to fine powders also comes under the general heading of nanotechnology, but we exclude observational techniques such as microscopy and various forms of surface analysis.

Nanotechnology is an "enabling" technology, in that it provides the basis for other technological developments, and it is also a "horizontal" or "cross-sectional" technology in that one technique may, with slight variations, be applicable in widely differing fields. A good example of this is thin-film technology, which is fundamental to electronics and optics. A wide range of materials are employed in devices such as computer and home entertainment peripherals, including magnetic disc reading heads, video cassette recorder spindles, optical disc stampers and ink jet nozzles. Optical and semiconductor components include laser gyroscope mirrors, diffraction gratings, X-ray optics, quantum-well devices.

参 考 文 献

[1] 李善术. 数控机床及应用 [M]. 北京：机械工业出版社，2001.
[2] 王宏文. 自动化专业英语教程 [M]. 北京：机械工业出版社，1998.
[3] 李伟光. 现代制造技术 [M]. 北京：机械工业出版社，2001.
[4] 卜令国，温艳. 计算机英语 [M]. 西安：西安电子科技大学出版社，2000.
[5] 秦荻辉. 科技英语 [M]. 西安：西安电子科技大学出版社，1999.
[6] 张燕，沈奇，付弘. 计算机英语 [M]. 西安：西安电子科技大学出版社，1999.
[7] 张学仁. 电火花线切割加工技术工人培训自学教材 [M]. 哈尔滨：哈尔滨工业大学出版社，2001.
[8] 将忠理. 机电与数控专业英语 [M]. 北京：机械工业出版社，2002.
[9] WARREN S. SEAMES. COMPUTER NUMERICAL CONTROL：CONCEPTS AND PROGRAMMING. NEW YORK. DELMAR PUBLISHERS INC，1990.
[10] Steve Krar, Arthur Gill. CNC TECHNOLOGY AND PROGRAMMING. Gregg Division：McGraw-Hill Publishing Company，1990.
[11] ALRN C DIXON, JAMES ANTONAKOS. DIGRAL ELECTRONICS WITH PICROPROCESSOR APPLICATIONS. USA：John Wiley & Sons，Inc，1987.
[12] 刘瑛. 罗学科. 数控技术应用专业英语 [M]. 北京：高等教育出版社，2004.
[13] 董建国. CAD/CAM 专业英语 [M]. 北京：机械工业出版社，2002.
[14] 黄毓瑜. 现代工业设计概论 [M]. 北京：化学工业出版社，2004.
[15] 詹姆斯 A 雷（James A Reph）亨利 W 克雷贝尔（Henry W Kraebber）. 计算机集成制造 [M]. 北京：机械工业出版社，2003.